T0213138

Managed Entry Agreements and Funding for Expensive Therapies

**Pharmaceuticals, Health Economics
and Market Access**
Series Editor: Mondher Toumi

Gene and Cell Therapies: Market Access and Funding
Eve Hanna and Mondher Toumi

Managed Entry Agreements and Funding for Expensive Therapies
Mondher Toumi and Szymon Jarosławski

For more information about this series, please visit: https://www.
crcpress.com/Pharmaceuticals-Health-Economics-and-Market-Access/
book-series/CRCPHEMA

Managed Entry Agreements and Funding for Expensive Therapies

Mondher Toumi

Szymon Jarosławski

CRC Press
Taylor & Francis Group
Boca Raton London New York

CRC Press is an imprint of the
Taylor & Francis Group, an **informa** business

First edition published 2022
by CRC Press
6000 Broken Sound Parkway NW, Suite 300, Boca Raton, FL 33487-2742

and by CRC Press
4 Park Square, Milton Park, Abingdon, Oxon, OX14 4RN

CRC Press is an imprint of Taylor & Francis Group, LLC

© 2022 Taylor & Francis Group, LLC

Reasonable efforts have been made to publish reliable data and information, but the author and publisher cannot assume responsibility for the validity of all materials or the consequences of their use. The authors and publishers have attempted to trace the copyright holders of all material reproduced in this publication and apologize to copyright holders if permission to publish in this form has not been obtained. If any copyright material has not been acknowledged please write and let us know so we may rectify in any future reprint.

Except as permitted under U.S. Copyright Law, no part of this book may be reprinted, reproduced, transmitted, or utilized in any form by any electronic, mechanical, or other means, now known or hereafter invented, including photocopying, microfilming, and recording, or in any information storage or retrieval system, without written permission from the publishers.

For permission to photocopy or use material electronically from this work, access www.copyright.com or contact the Copyright Clearance Center, Inc. (CCC), 222 Rosewood Drive, Danvers, MA 01923, 978-750-8400. For works that are not available on CCC please contact mpkbookspermissions@tandf.co.uk

Trademark notice: Product or corporate names may be trademarks or registered trademarks and are used only for identification and explanation without intent to infringe.

ISBN: 978-0-367-50029-0 (hbk)
ISBN: 978-0-367-50026-9 (pbk)
ISBN: 978-1-003-04856-5 (ebk)

DOI: 10.1201/9781003048565

Typeset in Garamond
by KnowledgeWorks Global Ltd.

Contents

CONTENTS

Series Preface

The major advances in the field of biotechnology and molecular biology in the twenty-first century have led to a better understanding of the pathophysiology of diseases. A new generation of biopharmaceuticals has emerged including a wide and heterogeneous range of innovative therapies. It aims to prevent or treat chronic or serious life-threatening diseases, which were previously considered as incurable. This unique book series focuses on how the regulatory environment has evolved to analyse and review those therapies while HTA agencies and payers remain resistant. It provides an insight into current learning by demonstrating how those products will be accessible and which policy changes will be required to permit patient access.

Authors

Mondher Toumi, MD, PhD, earned an MD by training, an MSc in biostatistics and in biological sciences (option pharmacology) and a PhD in economic sciences. He is Professor of Public Health at Aix-Marseille University. After working for 12 years as a Research Manager in the Department of Pharmacology at the University of Marseille, he joined the Public Health Department in 1993. In 1995 he was a carrier in the pharmaceutical industry for 13 years. Toumi was appointed Global Vice President at Lundbeck A/S in charge of health economics, outcome research, pricing, market access, epidemiology, risk management, governmental affairs and competitive intelligence. In 2008, he founded Creativ-Ceutical, an international consulting firm dedicated to support health industries and authorities in strategic decision making. In February 2009 he was appointed Professor at Lyon I University in the Department of Decision Sciences and Health Policies. That same year he was appointed Director of the Chair of Public Health and Market Access. He launched the first European University Diploma of Market Access (EMAUD), an international course already followed by almost 500 students. He recently created the Market Access Society to promote education, research and scientific activities at the interface of market access, HTA, public health and health economic assessment. He is editor in chief of the *Journal of Market Access and Health Policy* (JMAHP). In 2022, he founded

InovIntel a research and business venture specialised in artificial intelligence for life sciences. Mondher Toumi is also a Visiting Professor at Beijing University (Third Hospital). He is a recognised expert in health economics and an authority on market access and risk management. He has more than 250 scientific publications and communications and has authored or co-authored 15 books.

Szymon Jarosławski, MSc, PhD, earned his MSc in Biotechnology and PhD in Molecular Biology. He has over 15 years of experience conducting research and lecturing on Public Health and Entrepreneurship. His research areas include public policy, rare diseases, drug research and development and pricing, mental health, and counselling. Working as a consultant, he has offered decision-making advice to the life science industry and public healthcare authorities in Europe. He has also conducted research on Indian entrepreneurship, including medical technology, non-profit and social enterprise, telemedicine and tuberculosis diagnostic, as well as health systems. He has co-authored a few dozens of scientific articles and several academic books.

List of Abbreviations

AADC	Aromatic L-Amino Acid Decarboxylase
AAV	Adenovirus Associated Viral Vector
AB	Actual Benefit *(Service Médical Rendu)*
ACO	Affordable Care Organizations
ADA	Adenosine Deaminase Deficiency
AIFA	Italian Medicines Agency *(Agenzia Italiana del Farmaco)*
ALD	Adrenoleukodystrophy
ALL	Acute Lymphoblastic Leukaemia
ARM	Alliance for Regenerative Medicines
ASCT	Autologous Stem Cell Transplant
ATMP	Advanced Therapy Medicinal Products
ATU	Temporary authorisation for use *(Autorisation Temporaire d'Utilisation)*
AWU	Annual Work Unit
BCMA	B-Cell Maturation Antigen
BLA	Biologics License Application
CADTH	Canadian Agency for Drugs and Technologies
CAR	Chimeric Antigen Receptor
CAT	Committee for Advanced Therapies
CATP	Combined Therapy Medicinal Product
CBER	Centre for Biologics Evaluation and Research
CDF	Cancer Drugs Fund
CED	Coverage with Evidence Development
CEPS	*Comité Economique des Produits de Santé*

CFR	Code of Federal Regulations
CHMP	Committee for Medicinal Products for Human Use
CI	Confidence Interval
CMS	Centers for Medicare & Medicaid Services
COMP	Committee on Orphan Medicinal Products
CR	Complete Remission
CRISPR	Clustered Regularly Interspaced Short Palindromic Repeats
CRPC	Castrate Resistant Prostate Cancer
CTD	Common Technical Document
CTMP	Cell Therapy Medicinal Product
DLBCL	Diffuse Large B-Cell Lymphoma
DMD	Duchenne Muscular Dystrophy
DNA	Deoxyribonucleic Acid
DRG	Diagnosis Related Group
EC	European Commission
EMA	European Medicines Agency
ERP	External Reference Pricing
EU	European Union
FDA	Food and Drug Administration
GCP	Good Clinical Practice
GDP	Gross Domestic Product
GMP	Good Manufacturing Practices
GTMP	Gene Therapy Medicinal Product
GVHD	Graft Versus Host Disease
HAS	*Haute Autorité de Santé*
HCT/P	Human Cells, Tissues, or Cellular or Tissue-Based Products
HHS	Health and Human Services
HLA	Human Leukocyte Antigen
HPC	Hematopoietic Progenitor Cells
HRQoL	Health Related Quality of Life
HSCT	Hematopoietic Stem Cell Transplantation
HST	Highly Specialised Technology
HSV-TK	Herpes Simplex I Virus Thymidine Kinase

HTA	Health Technology Assessment
IAB	Improvement in Actual Benefit *(Amélioration du Service Médical Rendu)*
ICER	Incremental Cost-Effectiveness Ratio
ICER	Institute for Clinical and Economic Review
IM	Intramuscular
IND	Investigational New Drug
INN	International Non-proprietary Names
IP	Intellectual Property
IQWIG	*Institut Für Qualität Und Wirtschaftlichkeit Im Gesundheitswesen*
ITF	Innovation Task Force
ITT	Intention-To-Treat
IV	Intravenous
MA	Marketing Authorisation
MAA	Marketing Authorisation Application
MACI	Matrix-Induced Autologous Chondrocyte Implantation
MEA	Managed Entry Agreement
MPS	Mucopolysaccharidosis
NHL	Non-Hodgkin Lymphoma
NHS	National Health Services
NICE	National Institute for Health and Care Excellence
NIH	National Institutes of Health
OECD	Organisation for Economic Co-Operation and Development
OTAT	Office of Tissue and Advanced Therapies
PDCO	Paediatric Committee
PDUFA	Prescription Drug User Fee Act
PFU	Plaque-Forming Unit
PMBCL	Primary Mediastinal Large B-Cell Lymphoma
PRAC	Pharmacovigilance Risk Assessment Committee
PRIME	Priority Medicines
QALY	Quality Adjusted Life Year
RCT	Randomised Clinical Trial

RMAT	Regenerative Medicine Advanced Therapy
RMP	Risk Management Plan
SGCMPS	General Sub-Directorate of Quality of Medicines and Medical Devices
SMA	Spinal Muscular Atrophy
SMC	Scottish Medicine Consortium
SME	Micro-, Small- and Medium-Sized Enterprise
SPC	Summary of Product Characteristics
TC	Transparency Committee
TDT	Transfusion-Dependent β-Thalassaemia
TEP	Tissue Engineered Product
TLV	Dental and Pharmaceutical Benefits Agency
TPP	Target Product Profile
TTO	Time Trade-Off
UK	United Kingdom
US	United States
VBP	Value Based Pricing
WHO	World Health Organization
ZIN	*Zorginstituut Nederland*

Chapter 1

Introduction to Managed Entry Agreements

1.1 HISTORY

The arrival of expensive treatments (ETs) that have high therapeutic potential and address the highly unmet needs of patients may present extreme challenges to payers and manufacturers [1, 2]. The challenges arise from the high budget impact of these technologies for payers but also from the fact that they are oftentimes approved with limited evidence [3]. Indeed, there is a profound difference in how regulators and payers assess novel medicines; therefore, a positive marketing authorisation decision does not always translate into a positive reimbursement decision. Whereas regulators assess the benefit-risk ratio direction (positive, negative) of new technologies regardless of their cost and impact on the healthcare system, payers need to manage the magnitude of the added benefit versus standard of care or usual care and to put it in perspective with feasibility of use, the impact of the ET on the healthcare system, and in some countries the cost associated to the adoption of ETs. Therefore, the latter require an accurate appreciation of the incremental benefit of the ET versus the current treatment to be able to estimate payers' willingness to pay for the ET. If there is significant uncertainty regarding the benefit, and

DOI: 10.1201/9781003048565-1

eventually the cost, payers will not be able to make informed reimbursement decisions and they may deny financing of the ET or request a substantial price discount.

Also, institutionalised payers face increasing difficulty in denying positive reimbursement decisions due to discontent from patient associations and the general public [4] especially in poorly serviced populations, in disabling or life-threatening health conditions or when the expected benefit is believed to be very high. Further, when ETs are denied reimbursement by payers, manufacturers are unable to secure market access for products whose development may have consumed large sums of money. This may in turn disincentivise future research and development by the industry into therapeutic areas with significant unmet clinical needs.

Particularly in Europe, there have been cases of high-income countries denying financing of patient access to novel treatments because they were deemed too expensive. For instance, in the 2000s, Italy hesitated the public reimbursement of Alzheimer disease drugs and the UK has denied reimbursement by the National Health Service of several multiple sclerosis (MS) drugs and numerous cancer treatments [5–8]. To enable patient access to novel treatments and manage the financial burden of these technologies on public budgets, national payers and manufacturers have proven to be creative in proposing alternative reimbursement solutions. The so-called Managed Entry Agreements (MEAs) are contracts agreed between payers and industry to address the issue of high drug prices, especially if clinical evidence for the products is not well established [9–11]. Ever since, MEAs have been established in a range of very high-income countries, e.g., in Italy [12, 13], the UK [8], France [14], Australia, New Zealand [15], and Belgium, but also in medium- to high-income countries such as Poland and Hungary [16].

One of the first examples of MEAs in Europe was the CRONOS project that began in 2000 [17]. The Italian medicines agency (AIFA) developed this performance-based agreement to inform a definitive decision on the reimbursement of acetylcholinesterase inhibitors (ChEI) in Alzheimer's disease (AD) [18]. The drugs attracted AIFA's scrutiny because they assumed chronic treatment of a sizeable population of elderly patients. Moreover, it was suspected that only some patients may benefit from the therapy. Broadly, the rationale behind the CRONOS scheme was to evaluate if ChEI are effective in real-life clinical practice and to assess their long-term effectiveness and subsequently to make a reimbursement decision [19]. Consequently, CRONOS presumed collection and analysis of defined health outcomes from a cohort of AD patients in Italy. Such an approach to conditional reimbursement is known as coverage with evidence development (CED). It also assumed that the public health insurer reimbursed medicines only in patients who responded at four months of treatment, while the drug cost for non-responders was covered by manufacturers [20, 21]. Patients' individual data were reported in a specific registry. In 2004, the Italian agency concluded that CRONOS provided evidence that enabled the decision to fully finance these medicines, with restrictions regarding patient diagnosis and continuation of treatment and prescription limited to specialists [18]. This ended the course of this MEA.

In contrast to the Italian MEA, the UK's agreement for MS drugs has proven to be largely unsuccessful in terms of both patients access to the treatment, savings to payers and new evidence generation. Early on, the UK's HTA agency, the National Institute for Health and Care Excellence (NICE), considered that the incremental cost-effectiveness ratio (ICER) of β-interferons/glatiramer for the treatment of MS was too high for them to be financed by the NHS [22]. In 2002, the UK government established an MAE with four MS drugs manufacturers under which the drugs were to be financed by the NHS at a negotiated premium price, but the scheme assumed reduction of the price based if unfavourable effectiveness evidence from the scheme became available [23]. The MAE presumed also a follow up of 10,000 patients for over 10 years to assess whether the benefit observed in approval trials could be observed in practice [24]. This was a typical CED scheme. However, due to poor scheme governance and design flaws, no price reviews had taken place in the first 7 years of the scheme's operation [22]. Further, patients' access to the drugs was very limited. In 2007, there were only 11.5% of MS patients receiving the treatments in the UK vs. 35% in Western Europe and 50% in the US. While the MS scheme was developed as an experiment, the UK's government and NICE agreed on dozens of diverse, mostly financially-based MEAs in the following years [25, 26].

The above early examples show that while payers employed MAE as a way to manage uncertainty about the value of novel ET, manufacturers did so to prevent apparent reductions in the list prices of their products. Indeed, both the schemes assumed that the drugs will be sold at premium prices. The drive of manufacturers to preserve high list prices was dictated by the external reference pricing system that is used in many countries when pricing novel products. If prices are reduced in one country, it may have a spillover effect on other locations and lead to international price erosion.

Ever since, MAEs that obscure real-life net prices of novel expensive therapies have become widespread in Europe, but also all over the world. In the US, which features a fragmented market of private insurers alongside the public Medicare programme, apparent reduction of the list price for a single payer may lead to price erosion across remaining payers in the country and overseas. Therefore, MEAs have become an increasingly common tool perceived as a linkage between the value and the price of novel technologies in major pharmaceutical markets. However, while some of the MEAs aim at generating novel effectiveness evidence that complements data from approval trials, they are in most cases addressing affordability and willingness to pay of payers rather than uncertainty about drugs' value. They have become in most cases a disguised discount.

1.2 DEFINITION AND CONCEPTS

Over the last decade, there have been numerous approaches to classify the various types of MEAs and many terms have been proposed before the current terminology. MEAs have been known as innovative contracting, risk-sharing schemes, market access agreements, reimbursement schemes and patient access schemes to name a few [17, 27]. Interestingly, the term risk-sharing concept referred to the idea that payers and manufacturers should share the risk of novel therapies being potentially less effective in real clinical practice than in clinical trials submitted for marketing authorisation [28]. This principle was employed in both the Italian Cronos project and the UK's MS drugs scheme described above, with mixed results. However, most MEAs do not feature a risk-sharing mechanism, and, even if they limit payment for the drug in question to patients who achieve treatment success, they do not compare the real-life effectiveness against clinical trial efficacy in order to adjust the drug price. Therefore, the risk-sharing concept does not apply to the majority of agreements in place. In 2011, the Health Technology Assessment International (HTAi) Policy Forum coined the much broader term MEA, which has been later proposed as a preferred nomenclature by the International Society for Pharmacoeconomics and Outcomes Research (ISPOR) [9, 10, 29]. Three types of MEAs can be distinguished: finance-based agreements (FBAs) such as confidential discounts, performance-based agreements (PBAs) such as payment-by-results (P4P) or CED, and service-based agreements (SBAs), such as the patient support and care management solutions provided by the manufacturer [17, 27, 30–32]. The last has been recently described and labelled by M. Dabbous et al. [33]. More examples of each of the types of MEAs are discussed in a separate chapter.

1.3 GLOBAL PHENOMENON

Across geographies, the most common rationale of payers to introduce MEAs are budget impact constraints, followed by uncertainty about clinical outcomes or real-life outcomes. Manufacturers agree on MEAs to maintain high list prices of ETs.

Italy has the longest of experience in MEAs, and France, the Netherlands, Sweden, and the UK have a moderate level of experience. Also, Spain has some regional experience with the agreements. Canada and Germany have generally avoided MAEs. In the US, private insurers have the liberty to sign a variety of financial deals with manufacturers that typically involve discounts and occasionally payment-for-performance agreements.

AIFA has been the most prolific in Europe in terms of MEAs with over 100 agreements signed by 2018, however the UK has taken the lead more recently [34]. While individual discounts prevail in Italian MEAs they are followed by P4P deals cost-sharing and spending caps. Cost-sharing includes deals where the manufacturer funds the first treatment cycles, but the following cycles are covered by the national payer. CEDs are limited to early access schemes [8, 17].

Since 2006, AIFA has devoted several million euros to individual-patient PBAs for multiple costly medicines. Patients were enrolled into registries via online prescription forms completed by hospital specialists and tracked to monitor doctors' compliance and treatment outcomes [11]. In principle, they enable the national health system to only pay for patients who responded to treatment and ensure that specialist doctors prescribe the drugs to eligible patients only. However, the Italian MEAs allowed AIFA to recover only around 5% of the total expenditure on the drugs involved in the programs [11, 12]. Strikingly, simple discounts rather than PBAs were responsible for the majority of the savings [11, 12, 16–18]. This contrast with Sweden, where a 2018 review identified 56

MEAs that contributed to a striking 50% savings in the drugs cost to the national payer [35]. The authors concluded that the main driver behind MEAs in Sweden was the affordability rather than managing uncertainty.

In the UK, they were over 184 active MEAs by 2018. Seventy-two per cent were simple discounts, and 17% were commercial agreements, outcomes-based deals, or a combination of the two and mostly in oncology. Discounts are negotiated with manufacturers to achieve acceptable cost-effectiveness levels that is £20,000–30,000 per quality-adjusted life-year (QALY) gained and up to £50,000 for end-of-life treatments [36]. A much higher ICER range of £100,000–£300,000 was introduced in 2017 for ultra-orphan drugs assessed under the highly specialised technologies (HST) path [37].

In France, price-volume agreements are the source of most refunds paid by manufacturers to the national payers [38]. They accounted for 41% of the 1bn euro repaid by manufacturers to French authorities under the terms of all MEAs. Only 12% resulted from performance-based deals [38]. The preferred cost containment measure in France for orphan-designated products is the volume cap. In the vast majority of cases, the MEAs are used in France as a cost containment tool.

Overall, France and the UK implement MEAs to reduce the net cost of ETs while allowing companies to maintain high list prices. Further, Italian payer seems to only be willing to pay for ETs for patients who responded to treatment. Also, the Swedish and Dutch payers are inclined towards CED as a method of collecting new evidence to inform a final reimbursement decision. Increasingly, national payers in Central and Eastern Europe believe that MAEs may be a solution to a lower purchasing power of their economies and are willing to explore such agreements to mitigate budget impact of ETs.

The US payer environment is highly fragmented and each of the private insurers may engage in confidential negotiations with manufacturers that typically result in simple confidential discounts. One of the first outcome-based MEAs was a 1998 P4P scheme for simvastatin, where Merck would refund patients and insurers the cost of the drug if it did not help them lower low-density lipoprotein (LDL) cholesterol levels [39]. Frequently, the same P4P agreement can be signed by the company with multiple private insurers [40]. The public payer Centers for Medicare and Medicaid Services (CMS) is more restricted by regulations and also, it continues to favour MEAs for non-drug technologies.

The administrative burden of the MEA for physicians and payers is a key factor of consideration in France, Germany, Sweden, and the UK. Preferably, the efforts must be compensated by the generation of data that can be useful for the healthcare administration or the clinical community.

1.4 FROM MEA MANAGING PRICE TO NEW POLICIES MANAGING FUNDING

It is now understood that the MEAs manage uncertainty by reducing the net prices of ETs, but they do not necessarily address the budget impact of these technologies. The budget impact issue is particularly relevant for innovative health technologies, such as gene and cell therapies. Whereas such products come at an unprecedentedly high cost, they have the potential to cure patients and provide lifelong benefits with a single or short-term administration. However, some of the benefits of such ETs are only potential and not proven in long-term clinical trials; therefore, payers maintain that prices of certain products are not justified [41]. Therefore, payers are uncertain if the high upfront cost will be justified by deferred outcomes that are only potential. Hence, they may prefer to sign with manufacturers long-term instalment payment schedules that correspond to the outcomes horizon. Some propose that if a therapy is capable of extending patients' life or improving their quality of life beyond the treatment duration, then payment schemes for such therapies should reflect the extended benefits [42]. For instance, annuity or instalment payments allow payers to pay the cost of an ET based on an annual or some other schedule [43, 44]. Payment can also depend on certain criteria, such as patient outcomes. In this scenario, payers ensure a consistent cash flow that is distributed across the schedule but still record on their books the high upfront costs in the year the drug is purchased and administered to respect the generally accepted accounting rules. Noteworthy, drugs cannot be amortised or depreciated, because they are considered as consumables. Nevertheless, manufacturers' revenue is accounted in the year of the sale, while the cash flow is spread over the term of the payment schedule.

Further, in single-payer national healthcare systems, some government started to establish special silo funds to finance ETs that have shown well-proven efficacy [43]. Examples of such funds include the UK's Cancer Drug Fund, the Scottish "new medicines fund", and the Australian Complex Authority Required Highly Specialized Drugs Program. Further, in Italy, a "5% AIFA fund" is collected from the pharmaceutical industry and enables the financing of orphan ET. These funds are further described in Chapter 6. Manufacturers benefit from this model, because they are paid immediately at the time the payer purchases the therapeutic.

1.5 FUTURE PERSPECTIVE OF MEA

Overall, the most frequent motivation of payers to introduce MEAs are cost constraints, followed by uncertainty about real-life outcomes. Manufacturers propose MEAs in order to secure reimbursement while maintaining high list prices that are required by the industry for international price referencing.

However, outcome-based MEAs often involve extra resources and time use. This has to be balanced by a high unmet need or an important public health concern that is addressed by the medicine.

A special consideration is warranted for advanced medicinal products, such as gene therapies, that have shown significant health benefits in clinical trials but for which the maintenance of the benefits over longer time horizon cannot be ascertained. Whereas outcome-based MEAs, such as CED could in principle clarify this uncertainty, they are not feasible because patients would need to be followed for decades. Therefore, instalment payments are arising as a promising option for such ETs. However, the instalment payment will not address the actual payer challenge, which is the budget impact of ETs.

REFERENCES

1. Gronde TV, Uyl-de Groot CA, Pieters T. Addressing the Challenge of High-Priced Prescription Drugs in the Era of Precision Medicine: A Systematic Review of Drug Life Cycles, Therapeutic Drug Markets and Regulatory Frameworks. PLoS One 2017;12(8):e0182613.
2. Butcher A. Understanding Today's Drug Pricing Environment. Eur. Pharm. Rev. 2016;21(5):3.
3. Vella Bonanno P, et al. Adaptive Pathways: Possible Next Steps for Payers in Preparation for Their Potential Implementation. Front Pharmacol. 2017;8:497.
4. DiMasi JA. Price Trends for Prescription Pharmaceuticals: 1995–1999. 2000, Washington DC: Tufts University: Leavey Conference Center, Georgetown University.
5. Raftery J. Multiple Sclerosis Risk Sharing Scheme: A Costly Failure. BMJ 2010;340:c1672.
6. McCabe C, et al. Continuing the Multiple Sclerosis Risk Sharing Scheme Is Unjustified. BMJ 2010;340:c1786.
7. Sudlow CL, Counsell CE. Problems With UK government's Risk Sharing Scheme for Assessing Drugs for Multiple Sclerosis. BMJ 2003;326(7385):388–92.

8. Jaroslawski S, Toumi M. Design of Patient Access Schemes in the UK: Influence of Health Technology Assessment by the National Institute for Health and Clinical Excellence. Appl. Health Econ. Health Policy 2011;9(4):209–15.

9. Carlson JJ, et al. Linking Payment to Health Outcomes: A Taxonomy and Examination of Performance-Based Reimbursement Schemes between Healthcare Payers and Manufacturers. Health Policy 2010;96(3):179–90.

10. Grimm SE, et al. The HTA Risk Analysis Chart: Visualising the Need for and Potential Value of Managed Entry Agreements in Health Technology Assessment. Pharmacoeconomics 2017;35(12): 1287–96.

11. Vogler S, et al. How Can Pricing and Reimbursement Policies Improve Affordable Access to Medicines? Lessons Learned from European Countries. Appl. Health Econ. Health Policy 2017;15(3):307–21.

12. Villa F, et al. Determinants of Price Negotiations for New Drugs. The Experience of the Italian Medicines Agency. Health Policy 2019;123(6):595–600.

13. Tolley C, Palazzolo D. Managed Entry Agreements in UK, Italy and Spain. Value Health 2014;17(7):A449.

14. Sante CEP, Rapport D'Activite 2013. 2013: CEPS, HAS Online.

15. Babar ZU, et al., Patient Access to Medicines in Two Countries with Similar Health Systems and Differing Medicines Policies: Implications from a Comprehensive Literature Review. Res. Social Adm. Pharm. 2019;15(3):231–43.

16. Ferrario A, et al. The Implementation of Managed Entry Agreements in Central and Eastern Europe: Findings and Implications. Pharmacoeconomics 2017;35(12):1271–85.

17. Jaroslawski S, Toumi M. Market Access Agreements for Pharmaceuticals in Europe: Diversity of Approaches and Underlying Concepts. BMC Health Serv. Res. 2011;11:259.

18. AIFA. Progetto Cronos: i risultati dello studio osservazionale. 2004: Rome: Agenzia Italiana del Farmaco.

19. Lucchi E, et al. A Qualitative Analysis of the Mini Mental State Examination on Alzheimer's Disease Patients Treated With Cholinesterase Inhibitors. Arch. Gerontol. Geriatr. Suppl. 2004;(9):253–63.

20. AIFA. Protocollo di monitoraggio dei piani di trattamento farmacologico per la malattia di Alzheimer. 2000: Rome.
21. Adamski J, et al. Risk Sharing Arrangements for Pharmaceuticals: Potential Considerations and Recommendations for European Payers. BMC Health Serv. Res. 2010;10:153.
22. McCabe CJ, et al. Access with Evidence Development Schemes: A Framework for Description and Evaluation. Pharmacoeconomics 2010;28(2):143–52.
23. Chilcott J, et al. Modelling the Cost Effectiveness of Interferon Beta and Glatiramer Acetate in the Management of Multiple Sclerosis. Commentary: Evaluating Disease Modifying Treatments in Multiple Sclerosis. BMJ 2003;326(7388):522; discussion 522.
24. Health Service Circular HSC 2002/004. Cost Effective Provision of Disease Modifying Therapies for People With Multiple Sclerosis., D.o. Health, Editor. 2002, London: DoH.
25. NICE. List of patient access schemes approved as part of a NICE appraisal, 2010:
26. Towse A. Value Based Pricing, Research and Development, and Patient Access Schemes. Will the United Kingdom Get It Right or Wrong? Br. J. Clin. Pharmacol. 2010;70(3):360–6.
27. Garrison LP Jr., et al. Performance-Based Risk-Sharing Arrangements-Good Practices for Design, Implementation, and Evaluation: Report of the ISPOR Good Practices for Performance-Based Risk-Sharing Arrangements Task Force. Value Health 2013;16(5):703–19.
28. de Pouvourville G. Risk-Sharing Agreements for Innovative Drugs: A New Solution to Old Problems? Eur. J. Health Econ. 2006;7(3):155–7.
29. Klemp M, et al. What Principles Should Govern the Use of Managed Entry Agreements? Int. J. Technol. Assess Health Care 2011;27(1):77–83.
30. Bell K. First look at Simband, Samsung's health-tracking wearable of the future. 2014. cited 2019; Available from: http://mashable.com/2014/11/12/samsungs-simband
31. Ingelheim B. New Collaboration with Healthrageous on a digital diabetes self-management program. 2012, Boehringer Ingelheim Online.

32. Times F. FT Global Pharmaceuticals and Biotechnology Conference 2014. Available from: https://live.ft.com/Events/2014/FT-Global-Pharmaceuticals-and-Biotechnology-Conference-2014

33. Dabbous M, et al. Managed Entry Agreements: Policy Analysis From the European Perspective. Value Health 2020;23(4):425–33.

34. Wesley T. Market Access Trends in Europe. Market Access by Region/Europe. 2018, Datamonitor Healthcare. Informa.

35. Andersson E, et al. Risk Sharing in Managed Entry Agreements—A Review of the Swedish Experience. Health Policy 2020;124(4):404–410.

36. Bovenberg J, Penton H, Buyukkaramikli N. 10 Years of End-of-Life Criteria in the United Kingdom. Value Health 2021;24(5):691–98.

37. Pharmaphorum. Ultra-rare disease drugs: has access in England just got harder? 2017; Available from: https://pharmaphorum.com/views-and-analysis/ultra-rare-diseases-england/

38. CEPS. Rapport d'activité 2016. 2017; Available from: http://solidarites-sante.gouv.fr/IMG/pdf/rapport_annuel_2016_medicaments.pdf

39. Møldrup C. No Cure, No Pay. BMJ (Clinical research ed.) 2005;330(7502):1262–64.

40. Carlson JJ, Chen S, Garrison LP Jr. Performance-Based Risk-Sharing Arrangements: An Updated International Review. Pharmacoeconomics 2017;35(10):1063–72.

41. Kleinke J, McGee N. Breaking the Bank: Three Financing Models for Addressing the Drug Innovation Cost Crisis. Am. Health Drug Benefits 2015;8(3):118.

42. Cutler D, et al. Insurance Switching and Mismatch between the Costs and Benefits of New Technologies. Am. J. Manag. Care 2017; 23(12):750–57.

43. Rémuzat C, et al. Market Access Pathways for Cell Therapies in France. J. Mark Access Health Policy 2015;3.

44. Jørgensen J, Kefalas P. Annuity Payments can Increase Patient Access to Innovative Cell and Gene Therapies Under England's Net Budget Impact Test. J. Mark Access Health Policy 2017;5(1):1355203.

Chapter 2

Definition and Classification of MEAs

2.1 DEFINITION

Increasing budget constraints in health care and the emergence of expensive treatments (ETs) that have limited evidence at the time they are introduced to the market [1], present extreme challenges to payers and manufacturers [2, 3]. Also, payers face increasing difficulty in denying positive reimbursement decisions [4]. Since 2000s, the European Medicines Agency (EMA) has introduced new routes to market authorisation (MA) to expedite patient access, such as MA in exceptional circumstances, conditional MA, adaptive licensing, PRIME, and other routes [5–9]. As a result, drugs with insufficient evidence but high expected value entered the market. The so-called Managed Entry Agreements (MEAs) have been agreed between the institutionalised payers and industry to address payer concerns over funding ET that lack robust efficacy data [10–12]. Even though the ability of certain MEAs to achieve their goals is questionable [13–23], they have been established in a wide range of countries, e.g., in Italy [24, 25], United Kingdom (UK) [26], France [27], Australia, New Zealand [13], Belgium, Poland, and Hungary [28].

The World Health Organization (WHO) and the Organization for Economic Cooperation and Development (OECD) have both recognised the importance of MEAs

DOI: 10.1201/9781003048565-2

in the current pharmaceutical market, by creating a broad definition of the agreements:

> An arrangement between a manufacturer and payer/provider that enables access to (coverage/reimbursement of) a health technology subject to specified conditions. These arrangements can use a variety of mechanisms to address uncertainty about the performance of technologies or to manage the adoption of technologies in order to maximize their effective use, or limit their budget impact. [29]

2.2 TYPOLOGY OF MARKET ACCESS AGREEMENTS AND THEIR IMPACT ON THE PHARMACEUTICAL MARKET

Three types of MEAs can be distinguished: finance-based agreements (FBAs), performance-based agreements (PBAs), and service-based agreements (SBAs) [30–33]. Figure 2.1 presents a typology of MEAs [34].

2.2.1 Finance-Based Agreements

In an FBA, financing an ET is shared between the manufacturer and the payer. FBAs aim to contain ET costs and to facilitate the affordability of a product by various means that range from simple discounts to more complex utilisation caps and price-volume agreements [35]. In terms of utilisation caps, a payer may pay only for a specified number of patients in the country over a given period, with the remainder being paid for by the manufacturer. Alternatively, payer may pay only for a specified number of treatment cycles required by an individual patient, with the remainder of treatment being

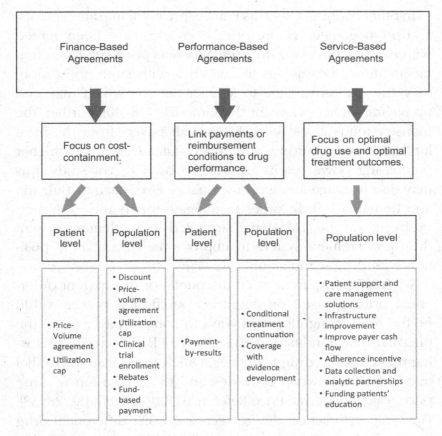

Figure 2.1 Typology of market access agreements.

covered by the manufacturer. However, simple rebates and discounts are the least burdensome FBAs and arguably provide the highest savings for the payers [36, 37].

Whereas cost-containment is the primary goal of FBAs, in countries that use the incremental cost-effectiveness ratio (ICER) to aid reimbursement decisions, such as the UK or Sweden, FBAs can be used to make costly products cost-effective. Importantly, under FBAs the ICER is only improved by reducing the cost and not by generating an additional health gain or new effectiveness data.

In other countries, such as France, price-volume agreements, and pre-negotiated future price decreases have been agreed with manufacturers for decades. Whereas price-volume agreements allow cost-savings to countries with large population, they cause disadvantage to smaller nations which may end up paying higher prices for the same ETs [38–40]. Further, the industry tends to establish FBAs with payers from the most lucrative, larger European Member States (EMS) with higher purchasing power parity. As shown by a recent study, this may disadvantage less wealthy, smaller EMS, such as Bulgaria and Romania [41]. In order to mitigate this equity distortion problem, several EMS have struggled to implement a joint drug procurement system to improve their negotiation position with drug manufactures [42–45].

Whereas transparent price discounts, or transparent differential pricing based on countries purchasing power would be the most straightforward ways to contain payer costs, the pharmaceutical industry uses secretive FBAs to avoid disclosing reductions to drug list prices, which could lead to parallel trade of ETs between countries and to the erosion of drug prices due to External Price Reference (ERP) system [12, 28, 38]. However, in practice, due to secretive FBAs differential drug pricing in Europe is already a reality, because countries that negotiated better FBA conditions end up paying lower net prices. Unfortunately, rather than reflecting the differences in the wealth among EMS, they reflect the differences in their negotiating power, placing smaller and less wealthy EMS in an underprivileged position. However, while a transparent differential pricing may be a fairer solution to address differences in affordability, it may also lead to parallel import and cause drug shortages in countries from which cheaper drugs would be exported [40, 46].

Most countries, with the exception of Italy, have tended towards FBAs [22]. In practice, MEAs have often a combination of traits characteristic to both, FBAs and PBAs described below.

2.2.2 Performance-Based Agreements

In contrast to FBAs, PBAs focus directly on the uncertainty around the effectiveness of a medicine by linking reimbursement decisions or drug payments to clinical outcomes/health gain produced by the product in clinical practice. Payers scrutiny over produced outcomes results from the fact that clinical trials may not always fully record patient-relevant outcomes, but focus on surrogate endpoints [47]. Also, if such trials are too short or feature single arm study designs, their results cannot be easily generalised by the payers into clinical practice in their healthcare systems.

Importantly, outcomes measured as a part of PBAs must be clearly defined as to their type and assessment time points. Further, the PBA itself needs to be subject to performance and quality benchmarks [30]. The novel data generated from PBA may be used by payers to reduce the uncertainties surrounding the effectiveness or the cost-effectiveness of the ET and its impact on the healthcare system.

, PBAs tend to be more difficult to implement than FBAs [14, 22]. Similar to FBAs, PBAs can be designed with respect to a population of treated patients or based on individual patients. The patient-level PBAs, also known as individual performance-based agreements (IPBAs), have proved to be challenging due to the high administrative burden of patient follow-up and the various confounding factors present in real-life clinical practice that may complicate statistical the analysis of effectiveness.

Interestingly, since 2006 Italy pioneered the development of IPBAs. The national medicines agency and Health Technology Assessment (HTA) body, Italian Medicines Agency (AIFA), has invested several million euros into a new system of real-world follow-up of outcomes of treatment with ETs that were reimbursed under dedicated IPBAs. However, due to the above-mentioned challenges, no rigorous effectiveness analyses for

ETs in the IPBAs have been published by the agency and also, AIFA recovered merely 5% of the total expenditure on the concerned ETs [17, 36].

Coverage with Evidence Development schemes (CEDs) are population-based PBAs under which payers agree to reimburse a therapy dependent on the manufacturer generating the missing evidence to address the uncertainty [48]. Upon CED completion, payers may adjust the conditions of drug reimbursement, including price or reimbursement rate alterations.

While most countries agree on state-of-the-art CEDs, the German Federal Joint Committee (Gemeinsamer Bundesausschuss, GBA) may routinely require the so-called time limited resolutions (TLR) for more than 20% of all early benefit assessments of ETs. However, in the majority of cases, TLRs correspond to additional data collection requirements that had been imposed already by the EMA at the drug's market launch [49].

An important shortcoming of CEDs is that payers still pay premium drug price during the evidence generation period, and they cannot recuperate the expenses even if the evidence does not support the claims of the manufacturer. For instance, between 2010 and 2016, £1.27 billion was made available to the UK NHS to cover ETs in oncology that were reimbursed as a part of CEDs. The so-called Cancer Drug Fund (CDF) was criticised for prioritising cancer drug expenditure while depriving investment in the whole cancer management pathway and for failing to deliver relevant outcomes [50, 51]. An improved, 2016 CDF was supposed to address these shortcomings by restricting the access to CEDs for products with immature evidence that had significant potential to be cost-effective [52]. While the Fund continues to consume £340 million annually, its value to the society remains unclear [52]. Further, channelling supplementary funds to a single disease area or to drugs with unproven effectiveness raises ethical concerns over patients with other types of diseases, who may be deprived access to more validated treatments or care, due to budget limitations.

In France, there is a single example of CEDs with an escrow agreement, where funds generated from ET sales were to be deposited in a public bank, and subsequently transferred either to the manufacturer or to the health insurance based on the outcome of the CED [30, 53, 54].

However, poor experience with CED is not limited to the UK. The Dutch National Healthcare Institute, Zorginstituut Nederland, was involved in a large number of CEDs [55, 56]. Unfortunately, the schemes' value to the payer has been questioned [57]. Illustratively, only one of the programs was completed within the stipulated time frame, and less than half of completed CEDs resulted in actionable recommendations. Strikingly, many of the schemes seemed to defeat their very purpose, which is evidence development for the payer. For instance, there were a number of inadequacies in the data collected and the schemes didn't take into account the initial recommendations of the payer [57]. Consequently, the Dutch payer shifted to FBAs.

The financing of ETs with immature clinical trial data is particularly problematic in the case of curative products that have the potential to provide long-term or life-long benefits with a single or short-term administration [58]. The long-term nature of such treatment outcomes would translate into CEDs with exceptionally lengthy, and possibly infeasible, time horizons [59]. To avoid such impractical CEDs, instalment payment that may be dependent on the sustained treatment response for each patient has been proposed [60]. However, the proposal of Avexis to split the $2 million payment for Zolgensma® into five instalments of $450,000, was not welcomed with enthusiasm by US payers [61]. In contrast, a payment-by-results agreement that linked drug payments to pre-established milestones has been agreed between Novartis and the Italian AIFA for the gene therapy Kymriah®.

Further, some experts referred to the use of the concept of amortisation where the payments for an ET would be spread over time during which the benefits of the health technology may be accrued by patients while the payments and costs of the disease are continuously discounted. These payments would occur according to a pre-defined schedule, and may be linked to certain disease-specific patient milestones. Shall patients' outcomes or milestones be not reached; payments would cease and further treatment costs would be transferred to the manufacturer [62–66]. However, since ETs are considered consumable goods, they would need to be considered as intangible assets to be amortisable according to generally accepted accounting rules.

Finally, whereas in principle, PBAs should focus on real-life, patient-relevant outcomes, in reality, such agreements in various countries are most frequently based on surrogate end-points because payers expect to be able to make a fast decision on drugs' performance [21, 40].

2.2.3 Service-Based Agreements

Services have historically been used as marketing tool of the pharmaceutical industry. However, more recently, there are services designed specifically for payers. SBAs are an emerging type of MEAs that include, e.g., the provision by the drug manufacturer of patient centres, patient or caregiver education, patient monitoring, real world data collection, etc., all related to the treatment of patients with the new, expensive product [67–73]. These services are meant to help payers address the effectiveness or usage uncertainty or enhance outcomes associated to an ET, patient care, and/or the health system needs or shortcomings. While SBAs are present in Europe, their origins can be traced back to the emergence of affordable care organizations (ACOs) that were incentivised by payers to deliver better patient outcomes. In this setting SBAs were meant to help ACOs achieve this goal via services ensuring the proper use of ETs, improving communication for payers, health care professionals, and patients, as well as data collection [74, 75]. Such services impact payers attitude when they are product-specific but are unlikely to have impact when they are product agnostic. A recent review proposed the following SBAs typology [34]:

1. Comprehensive disease management services, based on multi-stakeholder engagement.
2. Targeted services, such as performance-driven services, finance-driven services, and knowledge-driven services.
3. Performance-driven services, such as tools for enhancing compliance, remote patient monitoring, outcome monitoring, support healthcare provider decision, etc.
4. Finance-driven services, such as external or internal services or staff to improve healthcare providers' activities, patient management, etc.
5. Knowledge-driven services that aim to enhance healthcare providers', patients', or payers' insight.

2.3 CONCLUSIONS

Expensive technologies that are approved with immature efficacy data typically fail the traditional effectiveness and cost-effectiveness assessment criteria lied out by payers because of high uncertainty in extrapolation of effectiveness data. Since manufacturers are reluctant to reduce the prices and payers are often under political pressure to enable patients' access to the treatments, both parties agree on various MEAs. Shortly, the aim of FBAs is to reduce the net ET cost to payers; PBAs target the uncertainty about ET's real-life value to patients; and SBAs support the stakeholders in optimising the use of the ET. All types of MEAs enable manufacturers to maintain high list prices and confidential de-facto differential pricing.

Due to their simplicity, FBAs will likely remain the most common kind of MEAs. In some cases, it may be preferable to supplement or replace an FBA with an SBA, especially where there are unmet needs in infrastructure or stakeholder education. Unlike in the case of PBAs, the entire cost of running SBAs is typically born by the manufacturer, and so, there is no risk of the schemes becoming too burdensome for payers. Further, SBA have the potential to bring direct value to all stakeholders. However, as the conditional and accelerated approval of novel ETs becomes increasingly common, payers will continue to face significant uncertainty about drugs value for money. As such, they will insist on assessing drugs' real-life performance post-launch. Nevertheless, due to complex design, high administrative burden and uncertain value to payers, population-level PBAs such as CED will only continue to exist if they are relatively short-term. Individual-patient level PBAs may continue to be used in order to trigger instalment payments for potentially curative technologies with extremely high price tags.

REFERENCES

1. Vella Bonanno P, Ermisch M, Godman B, Martin AP, Van Den Bergh J, Bezmelnitsyna L, et al. Adaptive Pathways: Possible Next Steps for Payers in Preparation for Their Potential Implementation. Front Pharmacol. 2017;8:497.
2. Gronde TV, Uyl-de Groot CA, Pieters T. Addressing the Challenge of High-Priced Prescription Drugs in the Era of Precision Medicine: A Systematic Review of Drug Life Cycles, Therapeutic Drug Markets and Regulatory Frameworks. PloS One. 2017;12(8):e0182613.
3. Butcher A. Understanding Today's Drug Pricing Environment. Eur. Pharm. Rev. 2016;21(5):3.
4. DiMasi JA. Price Trends for Prescription Pharmaceuticals: 1995–1999. Leavey Conference Center, Georgetown University,

Washington DC: Tufts University, Department of Health and Human Services' Conference on Pharmaceutical Pricing Practices UaC; 2000 August 8–9, 2000. Available from: https://aspe.hhs. gov/price-trends-prescription-pharmaceuticals-1995-1999

5. Agency EM. Adaptive Pathways Online EMA. 2018. Available from: https://www.ema.europa.eu/en/human-regulatory/research-development/adaptive-pathways

6. Agency EM. Human Regulatory – Accelerated Assessment. 2018.

7. Agency EM. Human Regulatory – PRIME: Priority Medicines. 2019. Cited 2019. Available from: https://www.ema.europa.eu/en/human-regulatory/research-development/prime-priority-medicines

8. Lipska I, Hoekman J, McAuslane N, Leufkens HG, Hovels AM. Does Conditional Approval for New Oncology Drugs in Europe Lead to Differences in Health Technology Assessment Decisions? Clin. Pharmacol. Ther. 2015;98(5):489–91.

9. Bouvy JC, Sapede C, Garner S. Managed Entry Agreements for Pharmaceuticals in the Context of Adaptive Pathways in Europe. Front Pharmacol. 2018;9(280).

10. Carlson JJ, Sullivan SD, Garrison LP, Neumann PJ, Veenstra DL. Linking Payment to Health Outcomes: a Taxonomy and Examination of Performance-Based Reimbursement Schemes between Healthcare Payers and Manufacturers. Health Policy (Amsterdam, Netherlands). 2010;96(3):179–90.

11. Grimm SE, Strong M, Brennan A, Wailoo AJ. The HTA Risk Analysis Chart: Visualising the Need for and Potential Value of Managed Entry Agreements in Health Technology Assessment. PharmacoEconomics. 2017;35(12):1287–96.

12. Vogler S, Paris V, Ferrario A, Wirtz VJ, de Joncheere K, Schneider P, et al. How Can Pricing and Reimbursement Policies Improve Affordable Access to Medicines? Lessons Learned from European Countries. Appl. Health Econ. Health Policy. 2017;15(3):307–21.

13. Babar ZU, Gammie T, Seyfoddin A, Hasan SS, Curley LE. Patient Access to Medicines in Two Countries with Similar Health Systems and Differing Medicines Policies: Implications from a Comprehensive Literature Review. Res. Soc. Admin. Pharm. 2019;15(3):231–43.

14. Rubenfire A. Pay-for-Performance Drug Pricing: Drugmakers Asked to Eat Costs When Products Don't Deliver: Modern Healthcare; 2016. Available from: https://www.modernhealthcare.com/article/20161210/MAGAZINE/312109949/pay-for-performance-drug-pricing-drugmakers-asked-to-eat-costs-when-products-don-t-deliver

15. Toumi M. Introduction to market access for pharmaceuticals. 2017. Boca Raton: CRC Press.

16. Carlson JJ, Gries KS, Yeung K, Sullivan SD, Garrison LP Jr. Current Status and Trends in Performance-Based Risk-Sharing Arrangements between Healthcare Payers and Medical Product Manufacturers. Appl. Health Econ. Health Policy. 2014;12(3):231–8.

17. Navarria A, Drago V, Gozzo L, Longo L, Mansueto S, Pignataro G, et al. Do the Current Performance-Based Schemes in Italy Really Work? "Success fee": A Novel Measure for Cost-Containment of Drug Expenditure. Value Health: J. Int. Soc. Pharmacoecon. Outcomes Res. 2015;18(1):131–6.

18. Association DCotGM. Opinion on the "cost-sharing initiatives" and "risk-sharing agreements" between pharmaceutical manufacturers and health and hospital. 2008. Available from: http://www.akdae.de/Stellungnahmen/Weitere/20080508.pdf

19. Garrison LP Jr., Carlson JJ, Bajaj PS, Towse A, Neumann PJ, Sullivan SD, et al. Private Sector Risk-Sharing Agreements in the United States: Trends, Barriers, and Prospects. Am. J. Managed Care. 2015;21(9):632–40.

20. Faulkner SD, Lee M, Qin D, Morrell L, Xoxi E, Sammarco A, et al. Pricing and Reimbursement Experiences and Insights in the European Union and the United States: Lessons Learned to Approach Adaptive Payer Pathways. Clin. Pharmacol. Ther. 2016;100(6):730–42.

21. Toumi M, Jaroslawski S, Sawada T, Kornfeld A. The Use of Surrogate and Patient-Relevant Endpoints in Outcomes-Based Market Access Agreements: Current Debate. Appl. Health Econ. Health Policy. 2017;15(1):5–11.

22. Neumann PJ, Chambers JD, Simon F, Meckley LM. Risk-Sharing Arrangements That Link Payment for Drugs to Health Outcomes Are Proving Hard to Implement. Health Affairs (Project Hope). 2011;30(12):2329–37.

23. Toumi MMM. Define Access Agreements: Pharmaceutical Market Europe; 2011. Available from: www.pmlive.com/Europe

24. Villa F, Tutone M, Altamura G, Antignani S, Cangini A, Fortino I, et al. Determinants of Price Negotiations for New Drugs. The Experience of the Italian Medicines Agency. Health Policy (Amsterdam, Netherlands). 2019;123(6):595–600.

25. Tolley C, Palazzolo D. Managed Entry Agreements in UK, Italy And Spain. Value Health: J. Int. Soc. Pharmacoecon. Outcomes Res. 2014;17(7):A449.

26. Jaroslawski S, Toumi M. Design of Patient Access Schemes in the UK: Influence of Health Technology Assessment by the National Institute for Health and Clinical Excellence. Appl. Health Econ. Health Policy. 2011;9(4):209–15.

27. Sante CEdPd. Rapport D'Activite 2013. CEPS, HAS Online; 2013. Available from: https://solidarites-sante.gouv.fr/IMG/pdf/Rapport_d_activite_du_CEPS_en_2013_version_francaise_.pdf

28. Ferrario A, Araja D, Bochenek T, Catic T, Danko D, Dimitrova M, et al. The Implementation of Managed Entry Agreements in Central and Eastern Europe: Findings and Implications. PharmacoEconomics. 2017;35(12):1271–85.

29. Europe WHO. Access to new medicines in Europe: technical review of policy initiatives and opportunities for collaboration and research, 2015.

30. Garrison LP Jr., Towse A, Briggs A, de Pouvourville G, Grueger J, Mohr PE, et al. Performance-Based Risk-Sharing Arrangements-Good Practices for Design, Implementation, and Evaluation: Report of the ISPOR Good Practices for Performance-Based Risk-Sharing Arrangements Task Force. Value Health: J. Int. Soc. Pharmacoecon. Outcomes Res. 2013;16(5):703–19.

31. Bell K. First look at Simband, Samsung's health-tracking wearable of the future Mashable Mashable, 2014. Available from: http://mashable.com/2014/11/12/samsungs-simband

32. New Collaboration with Healthrageous on a digital diabetes self-management program [press release]. Boehringer Ingelheim Online: Boehringer Ingelheim Online, 2012.

33. Times F. FT global pharmaceuticals and biotechnology conference 2014 financial times live online. Available from: https://

live.ft.com/Events/2014/FT-Global-Pharmaceuticals-and-Biotechnology-Conference-2014

34. Dabbous M, Chachoua L, Caban A, Toumi M. Managed Entry Agreements: Policy Analysis from the European Perspective. Value Health. 2020;23(4):425–33.

35. Dunlop WCN, Staufer A, Levy P, Edwards GJ. Innovative Pharmaceutical Pricing Agreements in Five European Markets: A Survey of Stakeholder Attitudes and Experience. Health Policy (Amsterdam, Netherlands). 2018;122(5):528–32.

36. Garattini L, Curto A, van de Vooren K. Italian Risk-Sharing Agreements on Drugs: Are They Worthwhile? Eur. J. Health Econ. 2015;16(1):1–3.

37. Reuters. UK says Novartis, Bristol cancer drugs too costly Reuters Online: Reuters, 2010. Available from: https://www.reuters.com/article/britain-cancer/uk-says-novartis-bristol-cancer-drugs-too-costly-idUSLDE61718V20100209

38. Rémuzat C, Urbinati D, Mzoughi O, El Hammi E, Belgaied W, Toumi M. Overview of External Reference Pricing Systems in Europe. J. Market Access Health Policy. 2015;3:10.3402/jmahp.v3.27675.

39. Polimeni G, Isgrò V, Aiello A, D'Ausilio A, D'Addetta, Cuzzocrea S, et al. Role of Clinical Pharmacist in Optimizing Reimbursement Originating from Performance-Based Risk-Sharing Arrangements: The Experience of the University Hospital "G. Martino" from Messina, Italy. Value Health. 2016;19(7):A756. Available from: https://www.valueinhealthjournal.com/article/S1098-3015(16)33705-6/fulltext

40. Jarosławski S, Toumi M. Market Access Agreements for Pharmaceuticals in Europe: Diversity of Approaches and Underlying Concepts. BMC Health Serv. Res. 2011;11(1):259.

41. Young KE, Soussi I, Toumi M. The Perverse Impact of External Reference Pricing (ERP): A Comparison of Orphan Drugs Affordability in 12 European Countries. A Call for Policy Change. J. Market Access Health Policy 2017;5(1):1369817.

42. Mawdsley J. European Union Armaments Policy: Options for Small States? Eur. Security 2008;17(2–3):367–85.

43. 2019 IPC. Abstracts from the 4th International PPRI Conference 2019: Medicines Access Challenge – The Value of Pricing and Reimbursement Policies. J. Pharm. Policy Pract. 2019;12(1):34.

44. Wyszehradzką WZLPG. Joint purchases of medicines by the Visegrad Group. SEJM Online: SEJM – Polish Government, Parliament, 2017. Available from: http://www.sejm.gov.pl/sejm8.nsf/komunikat.xsp?documentId=4C3BC93DB93BC5CDC125820B00362FFE

45. Pisarczyk K, Rémuzat C, Hajjeji B, Toumi M. PCP8 – Beneluxa Initiative: Why the First Attempt of Joint Assessment Failed? Value Health 2018;21:S83.

46. Kanavos P. Differences in Costs of and Access to Pharmaceutical Products in the EU. In: European Parliament Online, editor. Policies DGFI, Policy PDAEAS, European Parliament, 2011. p. 92. Available from: http://www.europarl.europa.eu/RegData/etudes/etudes/join/2011/451481/IPOL-ENVI_ET(2011)451481_EN.pdf

47. De Sola-Morales O, Volmer T, Mantovani L. Perspectives to Mitigate Payer Uncertainty in Health Technology Assessment of Novel Oncology Drugs. J. Market Access Health Policy. 2019;7:1562861.

48. Garrison LP Jr., Carlson JJ, editor ISPOR short course 2014. ISPOR 2014; 2014: ISPOR.

49. Fortuna G. New drugs: how much are governments paying for innovation? EURACTIV Online: EURACTIV, 2018. Available from: https://www.euractiv.com/section/health-consumers/news/new-drugs-how-much-are-governments-paying-for-innovation/

50. Aggarwal A, Fojo T, Chamberlain C, Davis C, Sullivan R. Do Patient Access Schemes for High-Cost Cancer Drugs Deliver Value to Society?-Lessons from the NHS Cancer Drugs Fund. Ann. Oncol. Off. J. Eur. Soc. Med. Oncol. 2017;28(8):1738–50.

51. Boseley S. Cancer Drugs Fund condemned as expensive and ineffective Treatments approved by David Cameron's scheme were not worth money, extended life very little and often had adverse side-effects, study finds The Guardian Online: The Guardian, 2017. Available from: https://www.theguardian.com/science/2017/apr/28/cancer-drugs-fund-condemned-as-expensive-and-ineffective

52. Hawkes N. New Cancer Drugs Fund Keeps Within £340m a Year Budget. BMJ. 2018;360:k461.

53. Martelli N, van den Brink H, Borget I. New French Coverage with Evidence Development for Innovative Medical Devices:

Improvements and Unresolved Issues. Value Health J. Int. Soc. Pharm. Outcomes Res. 2016;19(1):17–9.

54. Whalen J. Europe's Drug Insurers Try Pay-for-Performance the Wall Street Journal Online: The Wall Street Journal, 2007. Available from: https://www.wsj.com/articles/SB119214458748556634

55. Nederland Z. Summary of recommendations by Zorginstituut Nederland dated 8 July 2016. Zorginstituut Nederland Online: Zorginstituut Nederland, 2016. Available from: https://english.zorginstituutnederland.nl/publications/reports/2016/07/08/fingolimod-gilenya

56. Nederland Z. Voorwaardelijke toelating/financiering van zorg. Zorginstituut Nederland Online: Zorginstituut Nederland, 2012. Available from: https://www.zorginstituutnederland.nl/over-ons/werkwijzen-en-procedures/adviseren-over-en-verduidelijken-van-het-basispakket-aan-zorg/beoordeling-voor-voorwaardelijke-toelating-van-zorg

57. Makady A, van Veelen A, de Boer A, Hillege JL, Klungel O, Goettsch W. Implementing Managed Entry Agreements in Practice: The Dutch Reality Check. Value Health. 2017;20(9):A702.

58. Remuzat C, Toumi M, Jorgensen J, Kefalas P. Market Access Pathways for Cell Therapies in France. J. Mark Access Health Policy. 2015;3.

59. Barlow JF, Yang M, Teagarden JR. Are Payers Ready, Willing, and Able to Provide Access to New Durable Gene Therapies? Value Health J. Int. Soc. Pharm. Outcomes Res. 2019;22(6):642–7.

60. Jorgensen J, Kefalas P. Annuity Payments can Increase Patient Access to Innovative Cell and Gene Therapies Under England's Net Budget Impact Test. J. Mark Access Health Policy. 2017;5(1):1355203.

61. Cohen J. At Over $2 Million Zolgensma Is The World's Most Expensive Therapy, Yet Relatively Cost-Effective Forbes Online: Forbes, 2019. Available from: https://www.forbes.com/sites/joshuacohen/2019/06/05/at-over-2-million-zolgensma-is-the-worlds-most-expensive-therapy-yet-relatively-cost-effective/#158d8d2c45f5

62. Alton GH, Calcoen D, Gottlieb S, Mcclellan MB, Mendelson D. How Will We Pay for the Cost of Cures? Opening Discussion:

Improving the Critical Path for Developing Cures. American Enterprise Institute Online: American Enterprise Institute, 2014. Available from: https://www.aei.org/wp-content/uploads/2014/07/-cost-of-cures_154738513625.pdf

63. Gottlieb SCT. Establishing new payment provisions for the high cost of curing disease. American Enterprise Institute Online: American Enterprise Institute; 2014 July 2014. Available from: https://pdfs.semanticscholar.org/4524/935a42720cfe57e28f1e53 9ef2066f909126.pdf

64. Vogler S, Paris V, Panteli D. European Observatory Policy Briefs. In: Richardson E, Palm W, Mossialos E, editors. Ensuring access to medicines: How to redesign pricing, reimbursement and procurement? Copenhagen (Denmark): European Observatory on Health Systems and Policies (c) World Health Organization 2018 (acting as the host organization for, and secretariat of, the European Observatory on Health Systems and Policies). 2018.

65. Kleinke JD, McGee N. Breaking the Bank: Three Financing Models for Addressing the Drug Innovation Cost Crisis. Am. Health Drug Benefits. 2015;8(3):118–26.

66. Hettle R, Corbett M, Hinde S, Hodgson R, Jones-Diette J, Woolacott N, et al. The Assessment and Appraisal of Regenerative Medicines and Cell Therapy Products: An Exploration of Methods for Review, Economic Evaluation and Appraisal. Health Technol. Assessment (Winchester, England). 2017;21(7):1–204.

67. AHIP. Issue Brief – Specialty Drugs: Issues and Challenges, 2015. Available from: https://www.ahip.org/Workarea/DownloadAsset.aspx?id=2147496213

68. Brook RACJ, Smeeding JE. Management of Specialty Drugs, Specialty Pharmacies and Biosimilars in the United States. J. Manag. Care Specialty Pharm. 2018;24(4-a):s101. Available from: http://tpg-nprt.com/portfolio-items/management-specialty-drugs-specialty-pharmacies-biosimilars-united-states/ #iLightbox[gallery732]/null

69. Brook RA. Specialty Pharmaceuticals and the Quest for Better Outcomes. J. Med. Econ. 2016;19(1):63–9.

70. Joachim AC, Kim J, Vranek K, Mattingly TJ II. Why Are Specialty Pharmaceuticals So Special? American Society of

Health-Systems Pharmacists Midyear Clinical Meeting, Las Vegas, Nevada, December 6, 2016.

71. Senior M. Digital Health Is Changing Health Care. Is it Changing Pharma? 2014. Available from: https://invivo.pharmaintelligence. informa.com/IV004232/Digital-Health-Is-Changing-Health-Care-Is-it-Changing-Pharma

72. Merck's WD. "Beyond The Pill" Bet, Vree Health, Goes Commercial, 2013. Available from: https://invivo.pharmaintell igence.informa.com/IV004113/Mercks-Beyond-The-Pill-Bet-Vree-Health-Goes-Commercial

73. B. D. Humana Bought Healthrageous to Build Out Vitality, 2013. Available from: http://mobihealthnews.com/26099/humanabought-healthrageous-to-build-out-vitality/

74. Stern D. Benefit Design Innovations to Manage Specialty Pharmaceuticals. J. Managed Care Pharm. 2008;14(4 Suppl): S12–6.

75. Patel BN, Audet PR. A Review of Approaches for the Management of Specialty Pharmaceuticals in the United States. PharmacoEconomics. 2014;32(11):1105–14.

Chapter 3

From Coverage with Evidence Development to Individual Performance-Based Agreements in Italy

3.1 CRONOS PROJECT

Italy has been an early adopter of performance-based agreements (PBAs). One of the best-known examples of PBAs was a national project called Project CRONOS that began in October 2000 [1]. The Italian medicines agency Agenzia Italiana del Farmaco (AIFA) developed this PBA to inform a definitive decision on the reimbursement of acetylcholinesterase inhibitors (ChEl) (donepezil, rivastigmine, and galantamine) in Alzheimer's disease (AD), which had some critics at that time [2]. AD is a progressive neurodegenerative disorder that results in deterioration of cognitive functioning, the appearance of behavioural/emotional troubles, and decreased ability to function independently. AD represents about 50% of all dementia cases and its prevalence increases rapidly in the population aged 65 and above [3].

Before the onset of CRONOS in Italy, several randomised, placebo-controlled clinical trials had consistently shown the benefit of ChEl in improving global

DOI: 10.1201/9781003048565-3

cognition in AD patients. The effect corresponded to a delay of 6–12 months in cognitive decline as compared to patients receiving placebo [4]. However, these studies used global cognition measures as outcomes and the cognitive performance across different domains remained unknown at the time market launch of ChEI [4]. Further, the effect on cognition of continuous treatment with ChEI beyond the duration of clinical trials was uncertain. Also, it was unclear if some patients may benefit more from the therapies and if the results of the clinical trials in a controlled environment would be generalisable to Italian clinical practice. The rationale behind the CRONOS scheme was to evaluate if ChEI are effective in real-life clinical practice, to collect more data on the qualitative aspects of the cognitive improvement resulting from the treatment, and to assess the effectiveness at a longer time horizon than the one used in clinical trials of these drugs [4].

As explained in an editorial letter of the scientific coordinators of CRONOS, "Pre-registrative trials (explanatory trials) are always conducted on the minimum number of patients and in the minimum time necessary to observe the outcome" and "it is mandatory for National Health Authorities to follow their use and effects in the real world" [5]. However, this perspective clearly ignores the fact that while post-marketing drug surveillance is meant to ensure that the benefit/risk ratio remains favourable when drugs are used in everyday practice, coverage with evidence development (CED) schemes are typically driven by high drug cost and/or significant budget impact of novel technologies. Therefore, as explained below, apart from collecting new outcomes and safety data, CRONOS also ensured that the Italian payer would not pay for patients who failed to respond within four months of treatment and also, it allowed the payer to reverse the reimbursement decision if the effectiveness results from the whole cohort of participating patients did not support the clinical trial outcomes.

To achieve these goals, CRONOS assumed collection and analysis of defined health outcomes from a cohort of AD patients. Importantly, special Units of Evaluation for Alzheimer were established for the evaluation of the outcomes of treatment with the drugs: after a four-month-run-in trial, and then every six months during the follow-up (Figure 3.1) [3]. It was carried out over two years in a nationally representative sample of 5462 patients with AD [2].

To participate in CRONOS projects, patients had to fulfil two criteria: obtain a score between 10 and 26 on the minimental state examination (MMSE); not be already treated with ChEl [6]. The patients underwent a comprehensive geriatric assessment, including an extended neuropsychological evaluation, brain computerized tomography (CT), and/or magnetic resonance imaging (MRI) scans (Figure 3.1) [6]. Global

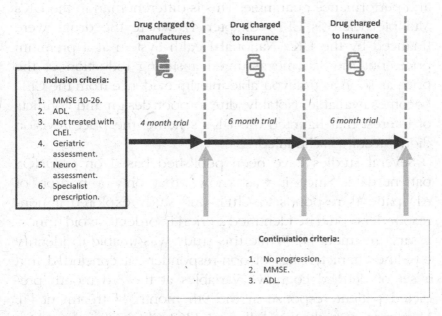

Figure 3.1 Summary the CRONOS process. The process is representing 2 six months cycle, but treatment is continued up to patient progression.

cognitive performance was assessed with MMSE, and global functional status was assessed with the activities of daily living (ADL) and the instrumental activities of daily living (IADL) at each evaluation [6].

As mentioned earlier, the agreement with manufacturers assumed that the public health insurer reimbursed medicines only in patients who responded at four months of treatment, while the cost for non-responders was covered by the manufacturers (Figure 3.1) [7, 8]. However, since the main objective of Italian scheme was to develop new evidence on real-life health outcomes from a cohort of patients that would enable a final reimbursement decision, it can be classified as CED and not as an individual performance-based agreement (IPBA). After the initial four-month trial period, patients were assessed every six months until disease progression.

Therefore, CRONOS was a mix of two approaches: CED and performance guarantee. This is different than in the UK's Multiple Sclerosis CED (Chapter 4), where the drugs were financed by the UK's National Health System at a premium price, but the scheme assumed a sliding reduction of the price as soon as unfavourable interim evidence from the CED becomes available. Notably, due to poor design and conflicts of interest that haunted the UK's scheme, the price revision didn't occur as scheduled.

Several studies have been published based on CRONOS patient data. Since it was known that only a portion of AD patients responds to ChEl, one study explored if non-responders could be identified a priori in order to avoid unnecessary treatment [4]. While this study was unable to identify baseline characteristics of non-responders, it concluded that a set of defined cognitive variables at the 3rd month predicted patient response at the 9th month of treatment [4]. Moreover, a study that followed 1362 CRONOS patients for nine months ascertained that the real-life benefits largely corresponded to those obtained in prior clinical trials of ChEI [9].

Further, a study that followed a sample of 66 AD patients over 21 months concluded, that while ChEI generated benefits for the first nine months of treatment, a progressive lowering of the MMSE scores was observed afterwards, at a rate of 0.9–1.0 points per year [3].

In 2004, the Italian AIFA agency concluded based on those analyses that CRONOS provided new real-life effectiveness data that justified the decision to reimburse these medicines, with some restrictions concerning diagnosis and continuation of treatment and prescription limited to specialist physicians [2].

However, in 2020, the CRONOS data from 938 patients was used to compare the relative effectiveness of the three ChEI used in the program, donepezil (Aricept©), galantamine (Reminyl©), and rivastigmine (Exelon©) over 36 weeks [10]. It found no significant difference in the effects of the drugs on cognitive or functional parameters assessed, but also concluded a lack of improvement in the MMSE and the Alzheimer's Disease Assessment Scale-cognitive subscale (ADAS-cog) scores at 36 weeks for all treatments. While the lack of efficacy might have been due to sub-optimal drug doses that had been used by patients, the study revealed that, at least at doses that were used in real-life practice, the drugs were ineffective at a longer time scale [10]. As ChEIs are products that target symptoms and not disease modifying therapies, it was not anticipated that a benefit could be seen in the long term but just at the time on treatment initiation. Even though patients continued to decline, the lack of an untreated, control population made it impossible to assess if disease progression was slower on ChEIs.

Other examples of CED in Italy were for drugs indicated in chronic angina pectoris (ivabradine) [11] and type 2 diabetes mellitus (sitagliptin, vilagliptin, and exenatide) [12, 13]. These MEAs aimed to monitor real-life use, collect epidemiologic data as well as new efficacy and safety data for re-assessment of price and/or reimbursement conditions for the medicines by the Italian agency. The schemes were run with a restriction of treatment initiation to specialist centres, monitoring of clinical practice, adverse events, and withdrawals due to treatment failure [14].

3.2 THE EMERGENCE OF IPBAs

It appears that the opportunity to only pay for patients that responded to treatment represented a greater value to Italian payer than the development of real-life effectiveness evidence that was enabled by the CRONOS project. Consequently, rather than investing in further CED, the Italian healthcare payer shifted their engagement to novel IPBAs. Therefore, they became an unavoidable path to market access for other innovative therapies.

Since 2006, AIFA has invested several million euros into IPBAs for multiple costly medicines. Patients were enrolled into registries via online prescription forms completed by hospital specialists and tracked to monitor doctors' compliance and treatment outcomes. Patient intake forms included eligibility according to the approved indication, evaluation of the effectiveness of treatment in clinical practice, epidemiological data, etc. [15].

By 2011, 19 various MEAs were operating in Italy: 12 for oncology drugs: erlotinib (2006), sunitinib (2006), sorafenib (in advanced renal cell carcinoma in 2006 and liver cancer in 2008), dasatinib (2007), bevacizumab (2008), lenalidomide (2008), temsirolimus (2008), bortezomib (2009), cetuximab (2009), lapatinib (2009), panitumumab (2009), and trabectedin (2009) [16]. The MEAs involved fixed discounts (from the list price) and/or paybacks for non-responding patients (100% or 50% of the drug's cost, all on a per-patient basis). While the safety and efficacy of the drugs were monitored in patient registries, it is noteworthy that those agreements did not seek to reduce uncertainty about a specified health outcome and data collection in the registries was not systematic with a high potential for various biases [13]. For example, in some regions of the country, only 50% of the patients were covered by the registries and patient selection was biased [17].

The costs of setting up and managing such registries remain underestimated [15]. However, despite significant investment of resources in the collection of real-life data, "no report published by AIFA has yet included relevant clinical outcomes" [15, 18, 19] from the registries. Nevertheless, they do enable the payer to only pay for patients who responded to treatment according to pre-specified criteria and ensure that specialist doctors prescribe the drugs in question only after thorough verification of patients' eligibility.

Whereas IPBAs aim at addressing the uncertainty about the drug's value for the payers, they often chose the rather uncertain, short-term surrogate endpoints as the primary endpoints in IPBA [20]. For example, the effectiveness of oncology drugs in Italian IPBAs was not measured based on their ability to extend patients life but rather based on surrogate markers, such as short-term disease-free progression of tumour biomarkers. In fact, 80% of patient registries in IPBAs are driven by surrogate endpoints [20]. One of our studies found that 85.3% IPBAs were based on surrogate endpoints, while only 4.6% of CED were based on such outcomes [20, 21].

Further, whereas IPBAs may achieve some cost-savings to the payer, they turned out to be disappointing. For instance, the Italian MEAs allowed AIFA to recover only around 5% of the total expenditure on the drugs involved in the programs [15, 18] versus 50% for Sweden [22]. However, simple discounts rather than IPBAs were responsible for the majority of the savings and were also cheaper to implement and report. Further, IPBAs were burdensome for the hospital staff. Consequently, doctors didn't always fill in the required patient forms and hospital administration failed to manage some claims [15, 18, 23–25].

3.3 CONCLUSIONS

CRONOS was an early and successful example of a CED scheme that provided novel clinical evidence which enabled final reimbursement and pricing decision-making. However, P4P had soon emerged as the preferred MEAs in Italy because they allowed the national payer to pay indefinitely only for patients who responded to treatment, without the need to analyse unbiased effectiveness data. Further, logistic challenges grown rapidly when such deals become ubiquitarian, creating a lot of biases in the information collected.

REFERENCES

1. Jaroslawski S, Toumi M. Market Access Agreements for Pharmaceuticals in Europe: Diversity of Approaches and Underlying Concepts. BMC Health Serv. Res. 2011;11:259.
2. AIFA. Progetto Cronos: i risultati dello studio osservazionale. 2004: Rome.
3. Fuschillo C, et al. Alzheimer's Disease and Acetylcholinesterase Inhibitor Agents: A Two-Year Longitudinal Study. Arch. Gerontol. Geriatr. Suppl. 2004(9):187–94.
4. Lucchi E, et al. A Qualitative Analysis of the Mini Mental State Examination on Alzheimer's Disease Patients Treated With Cholinesterase Inhibitors. Arch. Gerontol. Geriatr. Suppl. 2004(9):253–63.
5. Raschetti R, Maggini M, Vanacore N. Post-Marketing Studies: the Italian CRONOS Project. Int. J. Geriatr. Psychiatry 2003;18(10):962; author reply 963.
6. Mossello E, et al. Effectiveness and Safety of Cholinesterase Inhibitors in Elderly Subjects with Alzheimer's Disease: A "Real World" Study. Arch. Gerontol. Geriatr. Suppl. 2004(9):297–307.
7. AIFA. Protocollo di monitoraggio dei piani di trattamento farmacologico per la malattia di Alzheimer. 2000: Rome.
8. Adamski J, et al. Risk Sharing Arrangements for Pharmaceuticals: Potential Considerations and Recommendations for European Payers. BMC Health Serv. Res. 2010;10:153.

9. Bianchetti A, et al. Pharmacological Treatment of Alzheimer's Disease. Aging Clin. Exp. Res. 2006;18(2):158–62.

10. Santoro A, et al. Effects of Donepezil, Galantamine and Rivastigmine in 938 Italian Patients With Alzheimer's Disease: A Prospective, Observational Study. CNS Drugs 2010;24(2):163–76.

11. AIFA. La pratica clinica mette alla prova l'innovazione terapeutica: l'esempio "ivabradina". 2008: Rome.

12. AIFA. Incretine: il sistema di monitoraggio dell'AIFA. 2008.

13. Gallo PF, D. P. Pharmaceutical Risk-Sharing and Conditional Reimbursement in Italy. In: Central and Eastern European Society of Technology Assessment in Health Care (CEESTAHC), 2008.

14. AIFA. Criteri di valutazione per l'attribuzione del grado di innovazione terapeutica ai farmaci appartenenti a ciascuna delle tre classi della gravità della malattia bersaglio si considerano: la disponibilità di trattamenti preesistenti; l'entità dell'effetto terapeutico, 2007.

15. Garattini L, Curto A, van de Vooren K. Italian Risk-Sharing Agreements on Drugs: Are They Worthwhile? Eur. J. Health Econ. 2015;16(1):1–3.

16. AIFA. Oncology registries. 2010.

17. Jommi C. Central and regional policies affecting drugs market access in Italy. 2010: Bocconi University.

18. Navarria A, et al. Do the Current Performance-Based Schemes in Italy Really Work? "Success Fee": A Novel Measure for Cost-Containment of Drug Expenditure. Value Health 2015;18(1):131–6.

19. Garattini L, Curto A. Performance-Based Agreements in Italy: 'Trendy Outcomes' or Mere Illusions? PharmacoEconomics 2016;34(10):967–69.

20. Toumi M, et al. The Use of Surrogate and Patient-Relevant Endpoints in Outcomes-Based Market Access Agreements: Current Debate. Appl. Health Econ. Health Policy 2017;15(1):5–11.

21. Jarosławski S, Toumi M. Market Access Agreements for Pharmaceuticals in Europe: Diversity of Approaches and Underlying Concepts. BMC Health Serv. Res. 2011;11(1):259.

22. Andersson E, et al. Risk Sharing in Managed Entry Agreements—A Review of the Swedish Experience. Health Policy 2020;124(4):404–10.

23. Carlson JJ, et al. Current Status and Trends in Performance-Based Risk-Sharing Arrangements Between Healthcare Payers and Medical Product Manufacturers. Appl. Health Econ. Health Policy 2014;12(3):231–8.
24. Association, D.C.o.t.G.M. Opinion on the "cost-sharing initiatives" and "risk-sharing agreements" between pharmaceutical manufacturers and health and hospital, 2008.
25. Polimeni G, et al. Role of Clinical Pharmacist in Optimizing Reimbursement Originating from Performance-Based Risk-Sharing Arrangements: The Experience of the University Hospital "G. Martino" from Messina, Italy. Value Health 2016;19(7):A756.

Chapter 4

Coverage with Evidence Development for Multiple Sclerosis Drugs in the UK

A "Costly Failure"?

In response to negative drug financing recommendations by Health Technology Assessment (HTA) authorities, the industry has been negotiating Managed Entry Agreements (MEA). MEAs typically avoid simple reduction of the drug's list price, but instead, may attempt to link the drug's price to its real-life performance. Such performance may be as assessed on a population basis, e.g., in a patient registry. In this case, they are called Coverage with Evidence Development schemes (CED). This best-known early example of this approach is the 2002 scheme for MS disease-modifying drugs in the UK [1–3].

In August 1999, the UK Department of Health (DH) and the National Assembly for Wales (NAW) asked the National Institute for Health and Care Excellence (NICE) to appraise β-interferons/glatiramer for multiple sclerosis (MS) [4].

We will begin to report the story of the scheme by setting the context in which it was developed. In early 2000, representatives of the pharmaceutical and medical

DOI: 10.1201/9781003048565-4

43

device manufacturers asked NICE to mark all Provisional Appraisal Determination (PAD) and the remaining appraisal documentation as confidential material because disclosure to the public could have a significant impact on the manufacturers' share price and also on patient's confidence.

Despite that, in June 2000, a PAD on β-interferons leaked to the UK media [5]. NICE expressed disappointment that the confidentiality had not been respected. It turned out that one of the stakeholders (patients' groups, manufacturers, or health professionals) who had received the PAD for consultation allowed information to leak to the media.

In fact, NICE considered that "On the basis of a very careful consideration of the evidence, their [β-interferons] modest clinical benefit appears to be outweighed by their very high cost." Consequently, NICE advised that other than to patients already receiving these medicines, β-interferons should not be made available in the National Health Service (NHS) [4].

In its appraisal, NICE considered that the manufacturer models that estimated cost-utilities as costs per quality-adjusted life-year (QALY) widely differed [6]. Consequently, the estimates ranged from about £10,000 per QALY (a manufacturer's confidential estimate) to over £3 million per QALY (an American research group's findings). Further, the estimates were very sensitive to assumptions made in the modelling process, including the impact of a disease relapse on quality of life and the time horizon used. The NICE committee recognised that any inaccuracies in the data or methods used were liable to magnification in the extrapolation of 2–5 years benefit to 10–20 years or more.

The cost per QALY was assessed by NICE to range from £42,000 to £98,000 over 20 years period and would rise to as high as £780,000/QALY over a 5-year horizon [7]. Since the cost-effectiveness threshold was defined by NICE as a cost per QALY of no more than £35,000, the drugs were not recommended to be financed by the NHS [8]. Therefore, NICE refused to approve β-interferons to be used by the NHS for the treatment of MS on clinical and cost-effectiveness grounds.

In September 2000, an appeal to NICE's recommendation was submitted by: Association of British Neurologists ("ABN"), Biogen Limited ("Biogen"), Multiple Sclerosis Research Group ("MS Research Group"), Multiple Sclerosis Research Charitable Trust ("MS Research Trust"), Multiple Sclerosis Society ("MS Society"), The Neurological Alliance Royal College of Nursing ("RCN"), Schering Health Care Limited ("Schering") [9]. An Appeal Panel was set up by NICE that consisted of three non-executive directors of the Institute (who have not previously been involved in the appraisal) and two members nominated by patient organisations and the drugs' manufacturers.

The Panel upheld numerous points that were raised by the appellants [9]. Firstly, patient groups did not receive invitations to nominate a patient advocate to make representations to the Appraisal Committee. Therefore, the Appraisal Committee was not in a position to assess accurately the degree of clinical need of patients. Secondly, the Committee had not explained the basis of its conclusion that β-interferons are not cost-effective when compared to alternative uses of current resources. Appraisal Committee had not explained the basis of its conclusion that β-interferons are not cost-effective when compared with alternative uses of current resources. Thirdly, although the Panel agreed that Biogen's economic model wasn't acceptable, it concluded that the cited reasoning for rejecting the model had been perverse, implying that the model assumed an accumulation of benefit.

Further, the Panel concluded that the Appraisal Committee acted perversely in understating the potential clinical benefits of the treatments in the long term and that the decision to allow patients already on treatment to continue wasn't justified sufficiently. In addition, assessing the β-interferons in isolation, but not in comparison with the alternative use of NHS resources, making it impossible to ascertain on what basis the cost-effectiveness of β-interferons had been compared with alternative uses of NHS resources. Finally, the NICE guidance gave insufficient weight to the significance of magnetic resonance imaging (MRI) evidence on disease activity in early relapsing-remitting MS.

In the aftermath of the appeal, Schering submitted new modelling data which were rejected by NICE on methodological grounds. In December 2002, NICE commissioned the production of new economic models. As a result, several organisations were selected to produce the models: the Sheffield School of Health and Related Research (ScHARR) and the Department of Mathematics and Statistics at the University of Sheffield working with teams from Oxford City, Newcastle, Nottingham and York Universities [9].

Meanwhile, MS Research Trust provided new data from a survey that used a questionnaire sent to its members suffering from MS [10]. The questionnaire included the following items:

- type of MS, and symptoms,
- number of relapses,
- presence of difficulties in cognition,
- QoL,
- taking or not interferon or glatiramer.

Glatiramer (Copaxone®) was a new MS medicine that was granted marketing authorisation in the UK after the initial NICE recommendation was published [9].

The survey concluded that fatigue can profoundly disrupt the functioning of MS patients, and the possible benefits of disease-modifying drugs on fatigue warrant further study [10].

A consortium of universities commissioned by NICE prepared a report that implemented these results into their model [9]. It used data from patients with relapsing-remitting MS (RRMS) and secondary-progressive MS (SPMS) and a generalised linear regression model of EQ-5D single index score as a function of the type of MS, disability status, and relapse status to estimate the utility for each disability state and the disutility of relapse. However, the results on utility available to the NICE appraisal committee, were commercial-in-confidence.

The Expanded Disability Status Scale (EDDS) that was used in pivotal trials of the MS drugs features the following states:

- 0: normal neurological examination.
- 1.0–3.5: neurological impairments that are likely to have a limited impact on the activities of daily living.
- 4.0–5.5: ambulatory limitations for distances up to 500 m.
- 6.0-9.5: patient requires mobility aids. As the disease progresses, patients may require a wheelchair and eventually become bedridden.
- 10: death due to MS.

The difference between EDSS state 0 and state 3.0 was estimated to represent a 30% reduction in the patient's quality of life. There was a similar reduction in quality of life on declining from state 3.0 to state 7.0. Finally, the states 9.0 and 9.5 were worse than death that is, the quality of life was less than zero.

Based on the academic models, in August 2021, NICE published a draft decision where it recommended that DH engage in price negotiation with manufacturers [9]. However, no new prices were negotiated before the final appraisal was published in 2002. As a result, the new NICE report would issue a negative decision on the use of β-interferons and glatiramer for MS.

The patient decision aid (PDA) estimated that, over a 20-year time frame, the cost per QALY for the β-interferons and glatiramer may be at best of the order of £40,000–£90,000 [9]. These costs are much higher over a 10-year or 5-year time frame, displaying ranges of £190,000–£425,000 and £380,000–£780,000, respectively. In calculating cost per QALY using a 20-year time frame, the model had to assume that benefits accrue over a period much longer than is supported by any available evidence from clinical trials. Therefore, NICE concluded that there was an insufficient basis for this assumption.

Interestingly, in the academic model, EDSS scores were converted to changes in QALYs by applying utilities to each of the EDSS levels [9]. However, in the particular case of MS, although there are natural units that capture specific aspects of the impact of MS on patients, such as relapses avoided and delaying progression to wheelchair dependency, there are no outcomes that capture both the impact on relapses and the full impact on disease progression. Therefore, the model measures ignored some of the established benefits of β-interferons.

Consequently, a second appeal to the PDA is submitted in November 2001 by Teva Pharmaceuticals and Aventis Pharma Ltd jointly, Biogen Ltd and Schering Health Care jointly, Serono Pharmaceuticals Ltd, The Neurological Alliance, The Royal College of Nursing, The Multiple Sclerosis Society, and The Multiple Sclerosis Trust. However, none of the appeal points was upheld by NICE and the final negative recommendation is issued in February 2002 [9].

Nevertheless, between February and May 2002, DH engages with the manufacturers to develop an MEA to enable the financing of the drugs in the NHS [9]. In 2002, the government established a CED with four MS drugs manufacturers – Schering-Plough, Serono, Teva, and Biogen. The arrangement between the UK DH and the pharmaceutical companies was called MSRSS. The drugs were to be financed by the NHS at a premium, yet negotiated price, but the scheme assumed sliding reduction of the price based on the cost-effectiveness model used in the NICE appraisal if unfavourable evidence from the CED became available [11]. Importantly, only a designated neurologist would be able to prescribe these drugs to MS patients.

The CED also assumed a follow up of a cohort of approximately 10,000 patients for over 10 years to assess whether the benefit observed in phase III trials occurred in practice as well [12]. Patients were to be assessed using the Kurtzke EDSS, which was the same outcome measure as used in the trials. Importantly, EDSS is a subjective measure of disability composed of clinical judgement and patient self-reported elements [13]. However, due to the lack of better data sources, patient outcomes were supposed to be compared to a Canadian natural history cohort of MS patients from 1980, which was a questionable comparator.

The data was to be analysed every three years by an independent academic research group, the MSRSS Monitoring Study (MSRSS-MS). It was decided that if the cost per QALY over an envisaged 20-year modelling horizon was over £35,000/QALY, the price of drugs was to be reduced or refunds were to be given to the payer. Price reductions were assumed to occur as soon as interim data was available.

Effectively, the CED meant that both glatiramer and interferons were made available to all patients with relapsing-remitting MS, and those with secondary progressive MS, in whom relapses are the dominant clinical feature, and who meet the criteria developed by the Association of British Neurologists. It also meant that the cost of the drugs to the NHS would be significantly reduced (Table 4.1).

Table 4.1 NHS Risk-Sharing Scheme Agreed Prices

β – Interferons	Glatiramer
Avonex (Biogen): 8502 £/year of treatment (was 9061£: 6.2% reduction)	Copaxone (Teva/Aventis), 5823 £/year of treatment (was 6650£: 12.4% reduction)
βferon (Schering): 7260 £/year of treatment (was the same; no price reduction)	
Rebif (Serono) 7513 or 8942 £/year of treatment (was 9088£ or 12,068£: 17.3% reduction)	

4.1 PERFORMANCE OF THE MS CED

The MS CED scheme was heavily criticised for several reasons [14]:

- The model
  - Difficulties in fully assessing the quality of life and natural history of MS to the trial outcomes, which were based on changes in EDSS scores.
  - Assumptions about future discounting, not accounting for the cost of azathioprine.
  - Not accounting for early discontinuation of treatment by patients due to side-effects.
  - The subjectiveness of the EDSS scale that was used to assess patients in the study.
- The length of follow-up
  - Concerns that within the study horizon of ten years, there will be newer drugs available for MS patients.
- Funding and administration
  - Primary Care Trusts generally did not receive additional funding to fund the MS drugs.
  - Hospitals did not receive additional funding for more extensive follow-up consultations or for completing the necessary administrative forms.
  - Inadequacy of infrastructure that is required to manage the scheme, including specialist nurses.
  - Vested outcomes of members of the advisory group in the results of the study and the ban on public discussion about the scheme imposed on academics who were involved in the scheme.

In practice, the oversight of the MSRSS-MS was provided by a Scientific Advisory Group, composed of representatives of the four pharmaceutical companies, the UK Department of

Health Medicines and Industry Directorate, the MS Research Trust, independent academics, and the Director of the Academic team operating the data collection and analysis project [8]. Also, it appears that this body had the right to withhold permission for academics to publish results from the scheme and those were also forbidden to speak about the scheme without permission from the MS Research Trust [4]. The fact that the members of the scientific advisory group had interests in the continuation of the scheme, despite poor patient outcomes, raised concerns about the transparency of the scheme [15, 16]. Indeed, no price reviews had taken place in the first seven years of the scheme's operation [8].

Further, the MSRSS-MS was managed by the MS Research Trust, an organisation that appealed against the negative NICE recommendation. The MS Research Trust also owns the data collected by the scheme [4].

Early articles that were published regarding the outcomes of the MSRSS reported that "baseline characteristics and a small but statistically significant progression of disease were similar to those reported in previous pivotal studies" [17]. However, they also stated that it was too early to conclude on the cost-effectiveness of the MS disease-modifying drugs [4, 17].

Because the first published results of the evaluation of the scheme were much delayed and offered little evidence of cost-effectiveness, the House of Commons Health Committee suggested that it had been a "costly failure" to the NHS. One analysis recommended that financially based schemes are preferable to those based on outcomes [1].

James Gray MP, Chair of the All-Party Parliamentary Group on MS criticised the whole scheme, its methodology, inherent delays, and questionable conclusions. He said that it "had fatal flaws, is hopelessly out of date, has statistical and methodological inadequacies, its prescribing theory for the drugs is out of step with

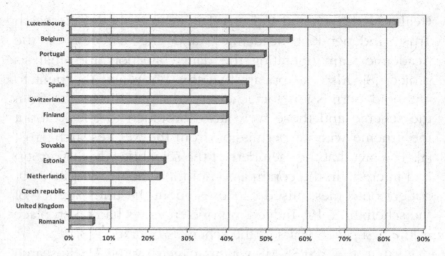

Figure 4.1 MS estimated proportion of patients on treatment in 2008.

medical science; the NICE guidelines on MS risk sharing drugs have stagnated; the scheme is knackered, it cannot be mended and anyone with half a brain would call for its abolition!"

Health Minister Mike O'Brien MP said that the government regarded the "treatment of MS as a priority for the NHS" but the recent two-year interim analysis "concluded that it was too early to reach firm conclusions about the cost-effectiveness of the drugs" and "the scientific advisory group is addressing several important methodological issues." These issues "would lead to more meaningful results when the next analysis takes place later this year."

The MEA was also a failure from the patients' perspective. In 2004, the uptake was very slow, 20% of eligible patients were still waiting to see a specialist doctor (Figure 4.1). Only 8% of MS patients received disease-modifying treatment. In 2007, there were 11.5% of MS patients receiving treatment in the UK vs. 35% in Western Europe and 50% in the US. Consequently, UK sales of MS drugs were the lowest in EU5 (Figure 4.2).

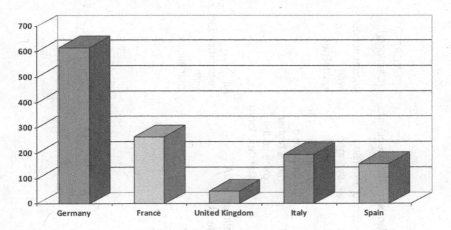

Figure 4.2   Sales of MS drugs affected by CED in UK.

While the MS scheme was developed as an experiment, since then NICE and DH have agreed on dozens of diverse, mostly financially-based MEAs [18, 19].

However, a 2015 clinical cohort study with natural history comparator confirmed the effectiveness and cost-effectiveness of interferon beta and glatiramer acetate in the UK MSRSS at six years [20, 21]. Nevertheless, a later systematic review and economic evaluation concluded that the MS drugs were cost-effective only for the treatment of clinically isolated syndrome, but not for relapsing-remitting MS [22].

4.2 HTA RECOMMENDATIONS FOR MS DRUGS IN OTHER COUNTRIES

UK remained isolated among western countries in its decision to hamper the reimbursement of MS drugs (Table 4.2). Even though the recommendations were sometimes restrictive and involved price reductions, they resulted in better patient access to MS drugs than that provided by the RSS in the UK.

Table 4.2 HTA Recommendations for MS Drugs across Countries

Drug	HAS (France)	CADTH (Canada)	INAMI (Belgium)	CVZ (Netherlands)	TLV (Sweden)	GENCAT (Catalonia)
Copaxone (Teva/Aventis)	Recommended only in RRMS (second line), SMR "important," ASMR "I," reimbursed at 65%	All recommended, MTA (MS drug comparison): Comparative evaluation is difficult. Estimate cost/QALY based on a two-year analytic horizon ranged from approximately $790,000 to $1,000,000	Recommended with restrictions, (The number of reimbursable packages is limited to a maximum 13 years)	Recommended only for RRMS,	Pre-filled syringes rejected (Manufacturer propose a 1.8 % higher price vs the standard form)	Recommended with guidelines
Avonex (Biogen)	Recommended (RRMS), SMR "important" ASMR "I," reimbursed at 65%	All recommended, MTA (MS drug comparison): Comparative evaluation is difficult. These drugs are incrementally more expensive for the improvement expected than many interventions currently used in various health conditions. Estimate cost/QALY based on a two-year analytic horizon for Copaxone®, Avonex®, and βseron® ranged from approximately $790,000 to $1,000,000	Recommended with restrictions (The number of reimbursable packages is limited to a maximum 13 years). Treatment cost/ patient /year = € 10,938	N/A Treatment cost/patient / year = €12,000	N/A	All recommended with guidelines

Table 4.2 (Continued) HTA Recommendations for MS Drugs across Countries

Drug	HAS (France)	CADTH (Canada)	INAMI (Belgium)	CVZ (Netherlands)	TLV (Sweden)	GENCAT (Catalonia)
βferon (Schering)	Recommended (RRMS, SPMS), SMR "important," ASMR "V" (versus Avonex), reimbursed at 65%		The company agrees a 1.4% price reduction. Treatment cost/patient/year = € 11,067	Treatment cost/patient/year = € 10,100	N/A	
Rebif (Serono)	Recommended (SPMS, RRMS) SMR "important," ASMR "I," reimbursed at 65%, (Rebif 22, the most cost effective)		Recommended with PAS – price reduction and a free pack per four weeks – treatment cost per patient per four weeks 1111 per four weeks amounts (15.5% lower than the first proposal)	N/A	Recommended, A new multidose form (Device) is recommended at the price of the pre-filled syringe	

4.3 CONCLUSION

It appears that there was a true uncertainty of the value of new MS drugs to the NHS and this led to a very wide ICER estimates for these therapies. The pressure from the public, media, patients, and health care professionals created conditions in which this CED was created. The decision to make MS treatments available on the NHS while new data was collected appears therefore justified. However, the agreement was flawed from the beginning, with little chance for the CED to deliver any relevant evidence. Moreover, the NHS deliberately prevented full patient access to these treatments by not appointing a sufficient number of neurologist to prescribe the drugs, thus containing their budget impact. In turn, manufacturers avoided to challenge the CED model as they continued to sell the drugs on other important markets outside the UK at premium prices. Unfortunately, many MS patients in the UK were at true loss as a result of the CED, as they expected access to new drugs but did not get it in reality.

REFERENCES

1. Raftery J. Multiple Sclerosis Risk Sharing Scheme: A Costly Failure. BMJ 2010;340:c1672.
2. McCabe C, Chilcott J, Claxton K, et al. Continuing the Multiple Sclerosis Risk Sharing Scheme Is Unjustified. BMJ 2010;340:c1786.
3. Sudlow CL, Counsell CE. Problems with UK government's Risk Sharing Scheme for Assessing Drugs for Multiple Sclerosis. BMJ 2003;326:388–92.
4. Pickin M, Cooper CL, Chater T, et al. The Multiple Sclerosis Risk Sharing Scheme Monitoring Study – Early Results and Lessons for the Future. BMC Neurol. 2009;9:1.
5. UK NICE provisional "no" to beta interferon/glatiramer on NHS for MS The Pharma Letter, 2000.
6. Excellence NIfC, Britain G. Beta interferon and glatiramer acetate for the treatment of multiple sclerosis. 2002: National Institute for Clinical Excellence.

7. Jaroslawski S, Toumi M. Market Access Agreements for Pharmaceuticals in Europe: Diversity of Approaches and Underlying Concepts. BMC Health Serv. Res. 2011;11:259.

8. McCabe CJ, Stafinski T, Edlin R, et al. Access with Evidence Development Schemes: A Framework for Description and Evaluation. Pharmacoeconomics 2010;28:143–52.

9. Multiple Technology Appraisal Beta interferon and glatiramer acetate for treating multiple sclerosis (review of TA32) [ID809] Committee papers. National Institute for Health and Care Excellence, 2017.

10. Hemmett L, Holmes J, Barnes M, et al. What Drives Quality of Life in Multiple Sclerosis? QJM 2004;97:671–6.

11. Chilcott J, McCabe C, Tappenden P, et al. Modelling the Cost Effectiveness of Interferon Beta and Glatiramer Acetate in the Management of Multiple Sclerosis. Commentary: Evaluating Disease Modifying Treatments in Multiple Sclerosis. BMJ 2003;326:522; discussion 22.

12. Health Service Circular HSC 2002/004. Cost effective provision of disease modifying therapies for people with multiple sclerosis. In: Health Do, ed. London: DoH, 2002.

13. Kurtzke JF. Rating Neurologic Impairment in Multiple Sclerosis: An Expanded Disability Status Scale (EDSS). Neurology 1983;33:1444–52.

14. Adamski J, Godman B, Ofierska-Sujkowska G, et al. Risk Sharing Arrangements for Pharmaceuticals: Potential Considerations and Recommendations for European Payers. BMC Health Serv. Res. 2010;10:153.

15. Gibson SG, Lemmens T. Niche Markets and Evidence Assessment in Transition: A Critical Review of Proposed Drug Reforms. Med. Law Rev. 2014;22:200–20.

16. Michelsen S, Nachi S, Van Dyck W, et al. Barriers and Opportunities for Implementation of Outcome-Based Spread Payments for High-Cost, One-Shot Curative Therapies. Front Pharmacol. 2020;11:594446.

17. Boggild M, Palace J, Barton P, et al. Multiple Sclerosis Risk Sharing Scheme: Two Year Results of Clinical Cohort Study With Historical Comparator. BMJ 2009;339:b4677.

18. NICE. List of patient access schemes approved as part of a NICE appraisal, 2010.

19. Towse A. Value Based Pricing, Research and Development, and Patient Access Schemes. Will the United Kingdom Get It Right or Wrong? Br. J. Clin. Pharmacol. 70:360–6.
20. Palace J, Duddy M, Bregenzer T, et al. Effectiveness and Cost-Effectiveness of Interferon Beta and Glatiramer Acetate in the UK Multiple Sclerosis Risk Sharing Scheme at 6 Years: A Clinical Cohort Study With Natural History Comparator. Lancet Neurol. 2015;14:497–505.
21. Palace J, Bregenzer T, Tremlett H, et al. UK Multiple Sclerosis Risk-Sharing Scheme: A New Natural History Dataset and an Improved Markov Model. BMJ Open 2014;4:e004073.
22. Melendez-Torres GJ, Auguste P, Armoiry X, et al. Clinical Effectiveness and Cost-Effectiveness of Beta-Interferon and Glatiramer Acetate for Treating Multiple Sclerosis: Systematic Review and Economic Evaluation. Health Technol. Assessment (Winchester, England). 2017;21:1–352.

Chapter 5

Country Comparison of the Implementation of Managed Entry Agreements

In this chapter, we review the implementation of Managed Entry Agreements (MEAs) in a selection of markets with a wide range of experience in innovative contracting. For each country, we discuss the past and current MEA trends and we illustrate those with representative examples of innovative contracts. Finally, we analyse the key hurdles and success factors relative to the implementation of MEAs across these jurisdictions, to guide companies and payers interested in developing agreements that suit their needs. We also attempt to discuss future trends in the field of innovative contracting.

5.1 FRANCE

Classical price-volume agreements (PVAs), rebates, or budget caps are the most common MEAs in France. The most prevalent innovative contracts include coverage with evidence development (CED) (e.g., a cohort study), although performance guarantee MEAs take place, especially in the oncology area. According to the French Economic Committee (CEPS), performance-based contracts refer to two situations: contracts based on individual clinical data or based on population data. The latter

DOI: 10.1201/9781003048565-5

situation is when conditional, premium drug price is linked to a need for a re-assessment of the improvement in actual benefit (ASMR) by the Transparency Committee (TC).

PVA are also the source of most refunds paid by manufacturers in France [1]. They accounted for 41% of the 1bn euro repaid by manufacturers to French authorities under the terms of all MEAs. Further, caps on daily per-patient treatment costs accounted for 9% and other finance-based MEAs for 38% of the repayments. As little as 12% resulted from performance-based deals [1].

As illustrated above, the French payers are reluctant to sign payment-for-performance (P4P) contracts. Only in situations where financial deal negotiations fail, a performance-based contract may be discussed. P4P contracts are a way to bridge the gap between the manufacturer price and the paucity of effectiveness data for expensive drugs with a rating of ASMR I-III.

The French payers are unlikely to want to replicate old models for innovative drugs that have a substantial effect size but only short-term efficacy data from clinical trials. These drugs certainly have to be evaluated in a longer timeframe, but the CEPS prefers financial-based contracts instead of long-term outcome-based ones.

In terms of successful MEA implementations, innovative products, preferably with a new mechanism of action, that demonstrate a large effect size within a short period are best suited for innovative contracting in France. Therefore, a minimum rating of ASMR III is usually required.

Further, the French payers are willing to enter innovative MEA if the product responds to a highly-unmet medical need and if the agreement addresses the uncertainty over the long-term effectiveness, cost-effectiveness, or budget impact. Also, innovative contracts have better chances of success when a small, specific target population is defined and when the MEA management is simple and lean. Nevertheless, the

French national industry association (Leem) has advocated for new approaches to HTA evaluation and pricing of innovative drugs that are more predictable and long-term oriented [2].

Finally, HAS in its cost effectiveness analysis guideline defines three situations when uncertainty about evidence is observed:

1. The company did not generate the right information while it was possible.
2. The company could not generate the right evidence and it is unlikely it will be feasible within a reasonable timeframe (three years).
3. The company could not generate the right evidence and it is feasible within a reasonable timeframe (three years).

In the first situation, HAS considers that a CED is not appropriate and the company is responsible for the lack of appropriate evidence. It is therefore to the company to generate the relevant evidence and come back to apply for reimbursement. In the second case, there is no reason to offer a CED and both authorities and payers should find an agreement on the value. In the third situation, a CED is an appropriate option.

5.1.1 CED/P4P for Actos® (Pioglitazone) and Avandia® (Rosiglitazone) in Type 2 Diabetes

In 2004, the CEPS demanded that the manufacturers of Actos and Avandia (Takeda and GSK) perform a two-year real-life study in type 2 diabetes to tackle the uncertainty around the real-life benefit of the treatments. The study investigated the delay in the switch to insulin. The premium drug price was conditional on the results of the study. ASMR re-assessment that took into consideration the results of the study did not show a higher benefit, and the drug prices were cut down [3–5].

5.1.2 P4P for Cimzia® (Certolizumab Pegol) in Rheumatoid Arthritis

To ensure that only responding patients are paid for, the French authorities concluded in 2013 a P4P agreement with UCB for Cimzia, in a highly competitive rheumatoid arthritis area. UCB agreed to refund the cost for patients who discontinued the treatment within the first three months. This deal allowed UCB to avoid price erosion that had been experienced by competitors as the company was allowed to maintain a premium list price [6, 7].

5.1.3 P4P for Imnovid® (Pomalidomide) in Multiple Myeloma

The P4P scheme, signed in 2015 between Celgene and CEPS for Imnovid in multiple myeloma, allowed the manufacturer to keep a high list price that was initially not accepted by the payers. Nevertheless, Celgene had to repay for every patient that did not respond to the treatment after three cycles of therapy. Celgene established a patient registry that was used to collect real-life safety and efficacy data which was analysed by CEPS yearly, to calculate due rebates [8, 9].

5.1.4 CED/P4P/PVA for Pradaxa® (Dabigatran), Xarelto® (Rivaroxaban) in Cardio- and Cerebrovascular Events

In 2012, French authorities raised concerns over the uncertainty related to real-life benefits in patient sub-populations and the budget impact caused by the extension of indication of Pradaxa and Xarelto. Therefore, they signed an outcome-based contract combined with a PVA for the drugs. Real-life data on cerebrovascular events in sub-populations were collected in the two main French patient records: "the reimbursed care in the outpatient setting" (SNIIRAM) and the "stays in public and private hospitals" (PMSI) databases. The results were re-evaluated by the Transparency Committee in 2016 who maintained the original ASMR rating of V for both drugs. As a result of the negative re-assessment, drugs prices were reduced by a confidential rebate [10, 11].

5.1.5 CED/P4P for Risperdal Consta® (Risperidone) in Schizophrenia

In 2005, Janssen and the French authorities signed a CED agreement for Risperdal Consta in schizophrenia. Janssen performed one-year real-life study on over 1800 patients that evaluated the number of hospitalisation due to acute schizophrenic episodes. During the study, the difference between the premium price and the price of generic comparators was deposited as public funds in the Caisse des Depots et Consignations. Depending on the study outcomes the money was to be transferred to Janssen or social security services The CED study demonstrated a 34% reduction in hospitalisation rate against other treatments. Positive real-life results supported drug additional benefit at an HTA re-assessment in 2010 and the initial premium price was maintained and the deposited funds were released to the manufacturer [4].

5.1.6 CED for Kymriah® (Tisagenlecleucel) in B-Cell Acute Lymphoblastic Leukaemia (ALL) and Diffuse Large B-Cell Lymphoma (DLBCL)

In 2019, the French authority HAS raised doubts over the efficacy, safety and complexity of treatment with Kymriah. It graded the drug's added benefit (ASMR) as important in ALL and minor in DLBCL. To resolve these doubts and enable yearly re-assessment of the ASMR, the agency requested a CED study based on data from several sources: the national CAR-T registry of the Lymphoma Academic Research Organisation (LYSARC), ongoing clinical trials and a post-authorisation efficacy study and from patients who received treatment through an early access scheme. Several data were collected, including key outcomes at 28 days, 100 days, 6 months, and then 6-monthly. In 2021, the agency maintained its original ASMR rating and drug reimbursement status [5, 12].

5.2 GERMANY

The majority of innovative contracts in Germany were concluded before the introduction of the Pharmaceuticals Market Reorganisation Act (AMNOG) in 2011. The law aimed to limit the increasing cost of pharmaceuticals and obliged manufacturers to subject their new products to an early benefit assessment by the Federal Joint Committee (G-BA) after being launched on the market. If no additional benefit vs standard of care is proven, the drug is allocated to a reference price group with comparable active ingredients. If there is no suitable reference price group, the umbrella payers association GKV-SV negotiates with the manufacturer on a refund rate which ensures no higher annual therapy costs than the comparative therapy. However, if an additional health benefit is proven, the GKV-SV negotiates with the manufacturer a premium on top of the price of the comparative therapy.

In Germany, one has to differentiate between MEAs concluded at the national and sickness-fund levels. Contracts at the national level, typically assume either flat pricing (mainly understood as cost per mg of drug) or PVAs – typically in the form of a price renegotiation if a threshold sales volume is reached. There are hardly any performance-based contracts at this level.

Further, at the sickness fund level, agreements include contracts that are signed within or after the initial period of 12 months during which drug prices are free and not controlled by the payer. These contracts can be signed in addition to deals agreed in AMNOG negotiations. The sickness funds typically only engage in MEAs if they can influence doctors prescribing and generate additional savings.

There are relatively few examples of performance-based contracts in Germany. However, the discussions around innovative contracting are very intensive, especially in the context of highly-priced advanced therapy medicinal products (ATMPs). Considering IQWIG's resistance towards real-world evidence collection, it is more likely that P4P will develop in favour of CED. There is also a debate on how to switch from upfront payment for contracted drugs to payment in instalments, especially if the drug's effectiveness could only be assessed after a long period. After the introduction of the high-cost pool, individual sickness funds are no longer interested in instalment payment agreements.

In terms of the kinds of treatment outcomes for innovative drugs, the German payers consider essentially two possibilities. They consider the new treatments to be either curative (no subsequent treatment is required) or having an intermediate health benefit (e.g., stopped disease progression). However, surrogate outcomes, such as progression-free survival (PFS) etc., are not accepted in the non-curative setting.

At the local level, the majority of sickness funds believe that it will be beneficial to act as an association and to have a contract at the national level as their position will be stronger. Interestingly, the German insurer Techniker Krankenkasse (TK) suggested introducing "dynamic evidence pricing" for ATMPs. The proposition includes regulated pricing for two years, followed by price negotiations based on collected data.

In terms of successful MEA implementations, there have been only two P4P contracts in Germany established after the initiation of AMNOG, and two more have been assessed by the arbitration board. The main focus of the payers is currently on MEAs for ATMPs.

In terms of possible hurdles to MEA development in Germany, the manufacturer may be required to build infrastructure for data collection for performance-based contracts. Further, the German medical association proposed that only a group of doctors with special contracts is included in MEAs and the doctors are also expecting an additional fee for extra work that will be required due to MEA management.

The German HTA agency IQWIG is very much attached to data from randomised clinical trials (RCTs), so data collection or evidence development post-marketing were unlikely elements of MEAs in this country. However, more recently, G-BA is requesting comparative data collection to assess relative effectiveness of highly innovative therapies (Chapter 7). Finally, based on previous MEA experiences, many German payers consider that the high administrative burden of MEA makes them unworthy of the savings they generated.

5.2.1 P4P and Discount for Kymriah® (Tisagenlecleucel) in B-Cell ALL and DLBCL

The 2019 P4P contract for Kymriah was the first MEA for an ATMP in Germany. The contract was signed between Novartis and GWQ ServicePlus, a purchasing body covering over 50 statutory health insurance companies that insure 8 million people. Under the terms of the agreement, the health fund will receive back part of the treatment costs for each patient who dies within a defined, but confidential, time. Further, the fund obtained a discount on the list price. The contract initially only applied until mid-September 2019. After the early benefit assessment, the reimbursement mechanism has reverted to a simple confidential discount [13].

5.2.2 P4P and Discount for Mavenclad® (Cladribine) in Multiple Sclerosis

In 2018, the G-BA decided, based on clinical trial data, that there is no added benefit of Mavenclad in multiple sclerosis. The real-life benefit was to be addressed by a P4P deal signed between Merck and GWQ ServicePlus. Merck would pay the cost of additional therapy required in years 3 and 4 of treatment. An additional 21% discount on the starting price was also negotiated. The agreement covered up to 6.5 million insured people. Interestingly, under the terms of the deal, the statutory health insurance would still make savings even if Merck's efficacy claims were sustained during the program because the annual cost of treatment with Mavenclad was well below that of comparable multiple sclerosis drugs [14].

5.3 ITALY

The Italian Medicines Agency (AIFA) has been the most prolific in Europe in terms of MEAs. As of 2018, AIFA agreed on 100 MEAs [15]. Currently, individual discounts prevail in Italian MEAs. They are followed by P4P deals (50% of all other agreements) and financial agreements, such as cost-sharing and spending caps. Cost-sharing includes deals where the manufacturer funds the first treatment cycles, but the following cycles are covered by the National Health Fund. CEDs are currently rare and apply mostly to early access schemes.

The shift towards discounts was driven by the perceived difficulty in the management of payment by results contracts. Further, between 2006 and 2012, manufacturers of drugs subject to CED and P4P in Italy had to rebate as little as 3.3% of the 3.7bn euro they earned from the sales of these drugs [17]. The figure is particularly disappointing to the Italian payer, given the high administrative burden of managing MEAs that are not simple discounts. Nevertheless, the number of P4P deals may continue to grow as the secrecy of discounts is increasingly problematic to the public. Indeed, at least seven cancer drugs and virtually all orphan drugs were subject to outcomes-based MEAs between 2017 and 2018 in Italy [15].

Further, spending cap deals have been decreasing in Italy because of the problems with estimating patient population sizes. Also, they sometimes resulted in manufacturers favouring to supply rich Italy regions that offered prescriptions prospects and speedy payments. However, overall, the Italian payers are open and well suited to test new MEA solutions for ATMPs.

When prescribing drugs that are included in registry-based MEAs, doctors in Italy must complete sophisticated intake forms that describe the disease and treatment outcomes and often involve in-depth patient assessment [15]. It is foreseen that more data will need to be collected, but also analysed, to serve physicians and hospitals. In the future, the outcomes data collected as a part of MEAs will be evaluated to assess the overall value of particular drugs and enable final reimbursement decisions that will put an end to these MEAs. Such analyses are currently not readily performed, despite data collection efforts.

In terms of future successful MEAs in Italy, they should be schemes that are easily manageable from an administrative perspective. For example, P4P contracts are more difficult to manage than discounts. If outcome-based MEAs are implemented, this should be preceded by an early dialogue between the regulators, HTA body, and manufacturer, and not merely be a result of post-launch considerations [16]. Further, therapeutic areas with a small patient population and with a short outcomes follow-up period (such as three to six months) are best suited for innovative contracts in e.g., oncology. An established concrete endpoint that gives payers a clear understanding of drug efficacy should be used to measure the treatment response. For example, in oncology, progression-free survival is commonly used and accepted by the Italian payers. Also, since the treatment response rate depends on the time of assessment, selecting an appropriate evaluation period is crucial. AIFA typically established that time based on the evidence from RCTs and the final decision on the evaluation period is negotiated with the manufacturer.

Further, drugs with high efficacy that was demonstrated in RCTs are more likely to be included in financial-based MEAs rather than in outcome-based deals. This is exemplified by six financial MEAs which AIFA signed for direct-acting antiviral products for chronic hepatitis C infections. Given the high efficacy of these drugs, P4P deals would be unlikely to bring about significant savings to the payer [15]. Notably, by the means of a PVA, AIFA was able to secure the lowest price in Europe for Sovaldi (sofosbuvir) – the first entrant in this therapeutic area.

In terms of known hurdles to MEAs in Italy, physicians do not have extra time allotted for the management of P4P contracts. In addition, AIFA's IT platform is not user-friendly and sometimes works slowly which adds to the burden. Due to the lack of time and platform inefficiencies, physicians sometimes fail to report all non-responders. This issue was partially addressed with a "success fee" deal for Esbriet described below, where the company was supposed to be paid for the drug only once the patients achieved treatment success.

Further, physicians are not able to retrieve any information from the AIFA registry. Giving access to collected data to hospitals would presumably encourage physicians to be more engaged in data collection. Also, the IT patient registries may not be fully functional at the time of product launch, resulting in data collection on paper forms which later need to be fed into the IT system. This creates inefficiencies and can create barriers to using the registry.

5.3.1 CED/P4P for Adcetris® (Brentuximab Vedotin) in Lymphoma

A mix of CED and P4P deals was used between 2012 and 2014 by AIFA to address the clinical uncertainty of Takeda's Adcetris in lymphoma and to grant early patient access. The agreement framework was close to the clinical trial design. Agreement terms included pay-back to the payer for each patient who progressed on the treatment or who had poor tolerance during the first four cycles. Further, the CED data collection was based on AIFA's Monitoring Registries System where a Drug Product Registry (DPR) was established for Adcetris for two years. The DPR tracked the eligibility of patients, monitored the appropriate use of the drug according to its indications and evaluated the real-life effectiveness of the treatment. Patients' epidemiology, drug safety profile, and ex-post evaluation data on any missing information could also be monitored. The registry data was then used by AIFA to negotiate a PVA with the manufacturer, which put an end to the CED/P4P scheme [17–19].

5.3.2 P4P for Aricept® (Donepezil) in Alzheimer Disease

In 2012, AIFA expressed concerns about the real-life durability of patient response to Pfizer's Aricept in Alzheimer disease. This issue was addressed with a P4P contract. Patient outcome data collection was based on AIFA's Monitoring Registries System. The authorities did not have to pay for non-responders, who did not demonstrate cognitive status improvement after six months. The P4P scheme allowed the drug to gain a positive reimbursement decision in mild-to-moderate Alzheimer's disease and a Class A reimbursement status [20, 21].

5.3.3 P4P and Indication-Specific Pricing for Avastin® (Bevacizumab) in Multiple Indications in Oncology

In 2011 and 2013 AIFA agreed on a number of MEAs to differentiate price and value across various oncology indications of Roche's Avastin. P4P schemes that used different rates of disease progression were set up for breast, colorectal, non-small-cell lung and renal cell cancers. Under the terms of the agreement, the payer would receive a partial or full payback of the drug cost for all patients who did not respond after six weeks of treatment. Moreover, Roche would provide continued treatment for free. Further, in the epithelial ovarian, fallopian tube or primary peritoneal cancers: Roche would pay for patients who permanently discontinued therapy due to disease progression or drug toxicity within the first eight months of treatment. As a result, while Avastin was reimbursed in seven oncology indications at a single list price, different P4P rules for each of the indications resulted in different de facto prices for each cancer type [22].

5.3.4 Portfolio Discount for a Combination of Herceptin® (Trastuzumab) and Perjeta® (Pertuzumab) in Breast Cancer

A portfolio discount was negotiated between AIFA and Roche for the combination of Herceptin and Perjeta, which enabled the reimbursement of Perjeta in Italy in 2013. Under the terms of the agreement, Roche gave a 50% discount for Herceptin when used in combination with Perjeta. A registry was set up by AIFA to monitor prescriptions and to manage the MEA. This MEA facilitated the establishment of Perjeta in clinical practice in Italy [23, 24].

5.3.5 Portfolio Agreement for Olysio® (Simeprevir) and Incivo® (Telaprevir) in HCV

To contain hepatitis C virus (HCV) drugs budget impact, in 2014 AIFA signed a portfolio agreement with Janssen-Cilag for two of its drugs marketed in the same indication. The MEA assumed de-listing of Incivo, a drug that did not reach the previously agreed 24-months cap, in exchange for the re-direction of payer savings to treat new patients with a novel HCV drug Olysio, at no additional cost for the payer. Importantly, patients who were already receiving Incivo could continue the treatment [25].

5.3.6 P4P for Strimvelis® (Autologous CD34+ Enriched Cell Fraction) in Severe Combined Immunodeficiency due to ADA

In 2016, following negotiations that secured a sizeable 50% list price discount, AIFA signed a P4P agreement for Strimvelis with GSK. AIFA aimed to minimise the risk of the unknown long-term benefit of this 594,000-euro therapy. The drug use was monitored through the AIFA's Monitoring Registries System, even though the treatment was only offered in a single hospital in Milan. GSK agreed to repay the Italian health service the full cost of treatment for each non-responding patient. However, due to disappointingly low Strimvelis sales of five doses, GSK vended the drug to Orchard Therapeutics and the MEA ended in 2018 [26–28]. According to the last AIFA update on MEA (05/04/2022), the agreement is currently applied.

5.3.7 P4P for Esbriet® (Pirfenidone) in Idiopathic Pulmonary Fibrosis

In 2013, AIFA signed a first-of-a-kind MEA with InterMune/Roche for its idiopathic pulmonary fibrosis drug Esbriet. Unlike in other P4P deals, the company supplied Esbriet to patients at no initial cost and was reimbursed only for Esbriet's use in patients that responded to the treatment after six months from the onset of therapy. Under the deal, a prescribing centre certified the successful or unsuccessful treatment outcome to the company, and a positive outcome certification triggered reimbursement by the payer. However, a failure in delivering the certificate to the manufacturer also resulted in reimbursement to the company, which led to inefficiencies and possible losses to the payer [17].

5.3.8 P4P for Kymriah® (Tisagenlecleucel) in B-Cell ALL and DLBCL

In 2019, AIFA signed a P4P deal for Kymriah over concerns with the drug's treatment outcomes. Patient data were collected and analysed by AIFA in its national web-based registry for medicines. Under this MEA, the payment by the public payer for each patient was conditional on their response to treatment and was split into three instalments that corresponded to patient follow-up points: at baseline, at 6 and 12 months. Data collection costs were borne by the manufacturer and analysis costs by AIFA [12].

5.4 NETHERLANDS

The most common MEAs in the Netherlands are financial agreements, including individual drug discounts, portfolio discounts, PVAs and capitated contracts. Financial arrangements are typically reserved for three circumstances: drugs with an above-average price or projected budget impact; drugs with unreasonable and unacceptable price tags; and drugs for which price competition is expected to be limited. The number of financial deals in the country has grown to 31 by 2020, producing 45.2% cost savings to the payer. Interestingly, public discounts exceeded €216 million, while confidential discounts €371 million [29].

Other types of contracts are rare and usually consist of CED. However, the Dutch payers are wary of CED agreements as they believe it is difficult to prove drug efficacy based on non-randomised observational studies. Overall, MEAs may be settled in a centralised way with the Ministry of Health or in a decentralised way with insurers or hospitals. Also, associations of insurers increasingly want to engage in MAE negotiations as they are not satisfied with those signed by the government.

There will likely be a shift towards P4P contracts as means of keeping the budget under control. However, it is possible that in the future Dutch model, the cost-effectiveness of groups of drugs, rather than of individual drugs, will be assessed by payers.

In terms of successful MEAs, products that feature a large effect size on well-established outcome endpoint, within a short period are best suited for MEA. These include products for cancers with short patient survival, certain gene therapies, as well as expensive products for small, well-defined patient populations, such as orphan drugs. However, the MEA type should be determined by motivations to enter the contract and not by product type or therapeutic area. For example, CED is a method to reduce the uncertainty around clinical trial results whereas P4P is a way to reduce uncertainty related to transferability and generalisability of these results into clinical practice. Also, the involvement of a strong, well-organised group of clinicians is a key factor for successful MEA in the Netherlands. In fact, drugs that are used routinely in clinical practice during the MEA period are unlikely to be removed from the reimbursement list. In contrast, drugs that do not become common practice are at the risk of being denied reimbursement in the future.

In terms of potential hurdles in MEA implementation, setting up a good patient registry is a complex and expensive endeavour. Therefore, it's preferable to connect to an existing registry. Nevertheless, patient enrolment in registries is often slow and poor and may not satisfy sample size requirements.

The quality of collected data is thoroughly scrutinised by payers. Also, a robust statistical analysis program will be required by payers beforehand.

Finally, financial measures need to fit within annual budgets, both at centralised and decentralised levels. Therefore, there is a preference to conclude simple financial deals that ensure up-front savings rather than delayed re-payments by manufacturers.

5.4.1 P4P and Discount for Imnovid® (Pomalidomide) in Multiple Myeloma

To avoid impact on hospital budgets, in 2016 the Dutch insurers decided with Celgene a P4P contract for Imnovid in multiple myeloma. Celgene agreed to refund the cost for non-responders. In addition, a price discount was negotiated for responders. The collaborative effort and strong backup of physicians were very important in the establishment and execution of this contract [30].

5.4.2 P4P for Kadcyla® (Trastuzumab Emtansine) in Breast Cancer

A P4P contract was concluded between the Dutch insurers and Roche to tackle the uncertainty of real-life benefits and the high price of Kadcyla in breast cancer. Under the terms of the agreement, Roche was to refund the cost of the treatment for non-responders, i.e., patients whose tumours failed to shrink in size within three months [31].

5.4.3 CED for Myozyme® (Alglucosidase Alfa) in Pompe Disease

Due to the high price (€400.000–€700.000 annually) of Myozyme for Pompe disease and the uncertainty of outcomes data from its clinical trials, the Dutch payer granted the drug a conditional reimbursement status for four years (2007–2011) [32]. The CED plan assumed that while the drug is reimbursed by the payer, the manufacturer runs several studies: an observational study designed to evaluate the usage, costs and outcomes of Myozyme; a prospective study in patients with the non-classical form of Pompe disease; a retrospective and prospective survey in patients with the non-classical form of Pompe disease and a prospective randomised Late-Onset Treatment Study (LOTS). Despite a negative cost-effectiveness re-assessment following this elaborate CED scheme, Myozyme was reimbursed for Pompe disease in the Netherlands due to high public pressure in 2013 [32].

5.4.4 CED for Replagal® (Agalsidase Alpha) in Fabry Disease

This CED for Replagal was similar to the deal from Myozyme described above. However, it assumed that the manufacturer runs two, partly overlapping cohort studies: a prospective study in patients with enzyme replacement therapy with a follow-up after six months and a retrospective comparative analysis between symptomatic patients who started enzyme replacement therapy before the onset of complications and a natural history cohort. Despite no cost-effectiveness proven by the CED scheme, Replagal retained reimbursement in Fabry disease in 2012 [32].

5.4.5 P4P for Xolair® (Omalizumab) in Asthma and Chronic Urticaria

In 2012, the Dutch payers signed a P4P deal with Novartis for its Xolair on the grounds of only a 70% response rate in treated patients. Under the deal, Novartis agreed to refund the cost of the first 16 weeks of treatment for non-responders. The response was assessed by a combination of outcomes agreed by pulmonologists and patients, including the number of exacerbations and urgent doctor visits within the first four months of treatment, patient score on the Asthma Control Questionnaire, Forced Expiratory Volume in one second, chronic steroid use, and Patient-Reported Outcomes. Given the extent of the data collected, pulmonologists required special training and extra time for patient assessment. Nevertheless, it was estimated that the P4P would generate €2 million of savings annually and so, the contract was renewed [33, 34].

5.4.6 CED for Spinraza® (Nusinersen) in Spinal Muscular Atrophy (SMA)

In 2020, the Dutch payer concluded that there wasn't sufficient data to assert if Spinraza was clinically and cost-effective in children over the age of 9.5 years, compared to no treatment. A seven-year CED study was commissioned to enable HTA re-assessment, versus historic registry data. A number of outcomes were to be collected in a registry developed for use by clinicians in the expert centre, with data points collected every three to six months for each patient. Every six months, data were evaluated by a research group from the centre. Data collection and analysis costs were borne by the manufacturer [12].

5.5 SPAIN

MEAs in Spain have mostly been agreed on the regional health level or with provider groups and individual hospitals. Catalonia has been one of the most active regions in this regard. By 2015, the Catalan health administration CatSalut had negotiated 18 MEAs for 10 drugs, mostly in oncology and rheumatology [35]. However, the regional leadership in MEAs has led to inequalities in access to new treatments across the country and national authorities are increasingly discussing national approaches to innovative contracting [15]. However, those are only foreseen for orphan and oncology drugs, that address a significant unmet medical need, are indicated for use in a small patient population, and with substantial budget impact [36, 37]. The situation changed in 2019 when the Ministry of Health started

Valtermed (https://www.sanidad.gob.es/profesionales/farmacia/valtermed/home.htm), the national register that lists the outcomes of different MEAs that are implemented at the national level. Today, even though regional authorities/hospitals can have their own MEAs, having a national outcomes database helps to centralize the process.

5.5.1 P4P for Iressa® (Gefitinib) in Non-Small Cell Lung Cancer

In 2011, the regional healthcare administration CatSalut and the Catalan Institute of Oncology (ICO), signed with AstraZeneca a pioneering P4P deal that allowed the treatment of 41 lung cancer patients with Iressa at the ICO. Patients were assessed for disease progression at eight and 16 weeks, with payments being linked to the treatment outcome [15, 38]. This approach was later scaled up to new hospitals and new treatments. Similar deals were signed in the following years for a few other oncology drugs, such as Avastin® (bevacizumab) in colorectal cancer and Erbitux® (cetuximab) in colorectal cancer.

5.5.2 P4P and Budget Cap for ChondroCelect® (Autologous Cartilage Cells) in Symptomatic Cartilage Defects of the Knee

In 2013, Tigenix' ChondroCelect entered into a P4P agreement with the national payer in Spain. Under the deal, the manufacturer would rebate treatment costs for high-level sportsmen who required a re-intervention for cartilage defects within three years. Moreover, the first treatment was delivered for free and the cost was capped at €20,000 for 200 patients. High complexity and relatively small public health importance led to the discontinuation of the scheme [39].

5.5.3 P4P for Cimzia® (Certolizumab) in Rheumatoid Arthritis

A 2013 P4P deal assumed that UCB funded treatment for patients who did not show improvement after 12 weeks of treatment with Cimzia. Despite previous successes of P4P contracts in Catalonia, the scheme was replaced with a discount due to complex management issues [40, 41].

5.5.4 Class Agreement and Budget and Utilisation Caps for Hepatitis C Drugs

In 2015, a national plan for hepatitis C drugs was developed to cover over half of the diagnosed population in Spain. This plan involved budget caps, utilisation caps, price-volume, and risk-sharing agreements. Communities were to receive up to €727M over three years to fund medicine. The government pled to secure the lowest prices of hepatitis C drugs in Europe [42].

5.5.5 P4P and Budget Cap for Sovaldi® (Sofosbuvir) in Hepatitis C

In 2014, national Spanish payers signed a P4P deal for Sovaldi with Gilead that also included a discount on the drug's price and a spending cap of €125m at year 1. The price discounts would be re-negotiated based on outcomes in clinical practice [42].

5.5.6 CED and Discount for Spinraza® (Nusinersen) in Spinal Muscular Atrophy (SMA)

In 2018, the Spanish ministry of health signed the first nationwide MEA for the orphan drug Spinraza with Biogen. The authorities were concerned that the clinical benefits of the drug had not been proven in patients with the most severe and mildest forms of SMA and also with the cure rates of only 50% in certain populations of children [37]. Under the terms of the deal, the annual per-patient cost of Spinraza would decrease from €400,000 in the first year of treatment to €200,000 yearly afterwards [37]. Further, the drug's effectiveness in the real world would be evaluated in a dedicated registry and this information would inform a final listing or de-listing decision [37].

5.5.7 CED for Dupixent® (Dupilumab) in Severe Atopic Dermatitis

In 2020, the Spanish ministry of health signed a nation-wide CED for Dupixent® for the treatment of severe atopic dermatitis. The drug was granted a premium price, but the financing was limited to patients who met a strict severity criteria. Further, doctors had to include each patient into a national Valtermed registry of MEAs, filling out the criteria for starting, monitoring, and discontinuing of the treatment, as well as patients' outcomes. Responders were defined as patients meeting a strict clinical criteria at 16, 24, and 52 weeks of treatment. Further, drug price revision was scheduled after one year or if monthly sales exceeded by more than 10% of the sales figure declared upfront by the manufacturer. Results of the national registry indicated that the real-life efficacy was above that from clinical trials and thus, the premium price was maintained [44, 45].

5.6 SWEDEN

In Sweden, MEAs are signed at the national level to prevent regional inequalities in access. However, regional county councils are also allowed to discuss the so called "side agreements" with manufacturers, but the initiative has to be on the manufacturer side. Prescribed drugs had a market of

34 billion SEK (approximately 3.4 billion Euros) in 2021. Side agreements are estimated to result in 2.7 billion SEK (about 8%) in payback. Here, we will find MEAs for many cancer drugs, e.g., Revlemide, Xtandi, Zytiga, but also MEA for TNFs, FVII, Factor-IX, and drugs for the treatment of hepatitis C.

In addition to MEAs in the market for prescribed drugs, there are also agreements in the market for requisition drugs (hospital drugs). Requisition drugs have a market of about 10 billion SEK (about 1.0 billion Euros). This market includes many cancer drugs, e.g., Keytruda, Enherto, Kymriah, Yescarta, but also infliximab for RA and Spinraza for SMA, and some ATMPs, e.g., Zolgensma, Luxturna [45]. Payers and administration of the hospitals generally do not see the need for innovative contracts, especially if they may result in extra administrative effort. The counties receive 60% of the rebates and central government 40%. In contrast to other countries, MEAs in Sweden often concern biosimilars and other therapeutic classes, not necessarily novel, innovative drugs [43].

In the years 2002–2010, several CED deals were signed by Swedish HTA agency Dental and Pharmaceutical Benefits Agency (TLV) to support final pricing and reimbursement. Currently, simple financial MEAs, such as discounts or refunds are the most frequent. Another kind of MEA that allows the payer to incur savings is a stopping rule – treatment duration is limited to a certain period or by a simple discontinuation criterion (e.g., disease progression). However, the arrival of new therapies with high up-front costs and significant uncertainty about disease recurrence, as well as drugs approved with single-arm studies, etc. will probably convince TLV to return to some kind of P4P deals. In terms of MEA success factors in Sweden, the contract type needs to be disease-specific and product-specific. For outcome-based deals, data should be collected for small patient populations. Sweden has a well-developed IT infrastructure for data collection. Registries are using unique Civil Personal Registration ID numbers that enable cross-matching across multiple databases. This potentially makes the implementation of the innovative contract easier.

In terms of potential hurdles, the administrative burden is the main barrier to outcome-based MEAs in Sweden. There are not enough nurses to collect the patient data and payers and hospitals are not willing to let in outsiders. Collection of data by patients was explored in pilot programmes that aimed at reducing admin burden on healthcare professionals, however, this triggered concerns over the security of data transfer.

Drug prices in Sweden are generally not relevant for international reference pricing, therefore companies do not need to struggle to maintain high list prices in this country. Therefore, the government considers abandoning confidential discounts in favour of open price negotiations. Further, since pharmacy and wholesaler margins are calculated based on list prices, this would allow producing additional savings on these margins.

5.6.1 CED for Duodopa® (Levodopa/Carbidopa) for Advanced Parkinson's Disease

In 2003, Neopharma's Duodopa was granted a premium price by TLV in Sweden on the condition that the company carries out a naturalistic CED study. However, the study, which was completed in 2005, was conducted without the TLV's approval and its results were not taken into consideration. The company undertook a new three-year, prospective health economic study and produced a formal economic model and was granted another two years to generate proper data. This data led TLV to conclude that the drug was not cost-effective and to order its delisting. The manufacturer run another study and improved the model which they re-submitted to TLV in 2008. This study allowed TLV to ascertain the drug's cost-effectiveness and confirm its reimbursement at the initial conditional price [47].

5.6.2 Financial-Based MEA and Data Collection for Entresto® (Sacubitril/Valsartan) in Heart Failure

In Sweden, Novartis and county councils were unable to ascertain the number of heart failure patients who were expected to be treated with Entresto. In 2016, Novartis agreed to submit follow-up data, including patient numbers, within two years. Data was collected in three registries and monitored by the TLV. Price of the drug was negotiated between the manufacturer and payers. This allowed better budget predictability [48].

5.6.3 Financial-Based MEA and Data Collection for Raxone® (Idebenone) in Leber's Hereditary Optic Neuropathy

Raxone in Leber's hereditary optic neuropathy was the first orphan drug available within an MEA in Sweden. The 2016 deal with Santhera resulted from uncertainties around the effectiveness and the number of patients who were expected to be treated with Raxone. Santhera agreed to submit usage data, including patient numbers, within two years. Under the agreement, the councils would receive paybacks from the company during the data collection period. This allowed better budget predictability [48].

5.6.4 Therapeutic Class Agreement for Hepatitis C Drugs (Direct-Acting Antivirals)

To improve budget predictability, a therapeutic class agreement was signed in 2015 between TLV and hepatitis C direct-acting antivirals manufacturers. Under the agreement, patient treatment was started with low-cost drugs and those with higher prices are subsidised only for patients who did not qualify for the low-cost drugs (i.e. more severe patients). In addition, discounts were granted in counties where many patients were treated or if treatment exceeded a pre-defined period [48–50].

5.6.5 Portfolio Agreement for Orkambi® (Lumacaftor/Ivacaftor) in Cystic Fibrosis

In 2018, TLV and counties signed a portfolio agreement deal with Vertex on the grounds of the lack of cost-effectiveness of Orkambi in cystic fibrosis. Under the deal, Vertex agreed to set one price for all of its cystic fibrosis treatments that would become available in a defined future. TLV agreed to reimburse the current future treatments of the company [51].

5.6.6 P4P/Financial-Based Agreement for Zytiga® (Abiraterone) in Prostate Cancer

After a negative reimbursement decision from TLV in 2013, Janssen signed with the Swedish association for county councils an MEA that involved a discount and pay-back for non-responders. In 2015, this deal was replaced by a national agreement with TLV, which assumed that the company would partially refund drug costs for initial treatment cycles and for cycles that exceeded an agreed treatment duration [48].

5.7 UK

In the past, the National Institute for Health and Care Excellence (NICE) tended to enter into complex agreements that were designed to address cost-effectiveness uncertainty [4, 52]. As of October 2018, NICE had 184 active, Patient Access Schemes (PAS). A total of 72% were simple discounts, 17% were commercial agreements, typically characterised by some form of conditionality, outcomes-based, or a combination of the two and pertaining mainly to oncology drugs. Other rare types of

MEAs included free stock (drug stock supplied at zero cost) and dose caps.

Discounts are negotiated with manufacturers to achieve acceptable cost-effectiveness levels. The incremental cost-effectiveness ratio (ICER) range that is generally deemed acceptable by NICE is £20,000–30,000 per quality-adjusted life-year (QALY) gained and up to £51,000/QALY gained if the severity burden of the disease can be shown for end-of-life treatments [53]. A much higher ICER range of £100,000–£300,000 was introduced in 2017 for ultra-orphan drugs assessed under the highly specialised technologies (HST) path [54].

The range of discounts negotiated with NICE was between 30% and 60%, with the majority of discounts within the 45%–50% range. Additional discounts may be applied by the NHS England if the drug's budget impact exceeds £20 million per annum. The level of discounts may be renegotiated following the collection of additional data, usually after three to five years.

Importantly, a positive NICE recommendation does not guarantee seamless market access in the UK. This applies even to highly effective treatments whose ICER is within the range that is accepted by NICE. This was best exemplified in the case of Gilead's Sovaldi (sofosbuvir) for chronic hepatitis C, which was accepted by NICE but was not readily available through the NHS because of its significant budget impact [55]. Many clinicians considered that the NHS England was intentionally stalling patient access to the drug, which patient support groups threatened the public payer with legal action [55]. It took nearly two years since the initial positive NICE appraisal in 2015 for the NHS to eventually strike a P4P deal with hepatitis C drug manufacturers that boosted patients' access. Under the terms of the agreement, the NHS would be refunded for the cost of drugs prescribed to patients who failed to respond despite completing the treatment course [56].

However, CED are also present in the UK – as a part HST assessment procedure, as well as within the Cancer Drug Fund

(CDF) evaluations [57, 58]. The most common reasons for NICE to sign such schemes is lack of "proven" cost-effectiveness, due to the uncertain clinical or other data for the economic analysis, uncertainty related to relative effectiveness and the need to collect more evidence on long-term outcomes and adverse effects of treatment [57]. For HST drugs, its common to have a financial-based MEA involving CED, as cost-effectiveness is always uncertain due to data limitations in rare diseases. CED conditions specify areas of clinical uncertainty and the methods through which new data will be produced. After a specified data period NICE undertakes a second assessment to reach a final recommendation for the NHS. CED deals negotiated as a part of the CDF feature a fixed budget and a rebating mechanism that ensures that any overspend is covered by manufacturers. On top of the CDF negotiations, NHS England is also engaging in confidential discount discussions with manufacturers that lead to further cost-savings [15]. The innovative medicines fund is a new initiative in England and Wales to operate like the CDF but to cover innovative non-cancer drugs recommended by NICE.

Further, the government-commissioned Accelerated Access Review (AAR) (October 2016) proposed a broader approach to using MEAs. UK payers are currently working on novel ways to price ATMPs. Delayed payments are broadly discussed; however, the exact approach is not known yet. The Cancer Research UK issued a report in which it calls for an outcomes-based payment system for new cancer drugs.

In terms of successful outcome-based MEA, they can only be concluded for drugs that address high unmet clinical need. The contract has to address the kind of cost-effectiveness uncertainty that can be resolved by a well-designed study. However, the scheme must not require excessive infrastructure and human resources. Performance-based schemes should be based on objective outcome measures that are generated by treatment.

In terms of potential hurdles, NICE has a very strong preference for simple price discounts. However, failure to propose a discount early in the assessment stage may lead to a negative reimbursement recommendation. This may take months to be reverted, even if the manufacturer is willing to offer a price reduction. This was illustrated by the struggle of Pfizer to secure the UK's market for its breast cancer drug Ibrance (palbociclib), which experienced nearly one year of delay in accessing the UK's market [59]. But this could have been avoided, had the company offered a discount early on.

In terms of outcomes-based MEAs, there is a recognised need for better data infrastructure in the NHS. The cancer data infrastructure in place is capable of capturing some of the main treatment outcomes, such as survival or disease progression. However, there is a need to envision to what extent data on other outcomes of importance, such as long-term side effects and return to normal activities, can be collected. Importantly, outcomes-based MEAs will require data collection, sharing and analysis by the NHS staff, which is generally not welcomed by the UK's authorities. MEAs may also be implemented to encourage drug research and development in neglected therapeutic areas. The new subscription-like model named "Netflix payment" is being implemented by the UK's NHS for novel antibiotics for resistant infections – Pfizer's Zavicefta (ceftazidime/avibactam) and Shionogi's Fetcroja (cefiderocol). Under the terms of the 10-years agreement, NHS England will pay a fixed annual fee of £10 million per year for access to the two medicines, irrespective of how much they are used to treat patients.

5.7.1 CED for Kymriah® (Tisagenlecleucel) in B-Cell ALL and DLBCL

Kymriah® was assessed as a part of the Cancer Drugs Fund for B-cell ALL and DLBCL in 2018 and 2019, respectively. The appraisal pointed to clinical uncertainties and data paucity which had to be addressed by further data collection for four years and seven months in ALL and four years in DLBCL indications. The CED was informed by several data sources: ongoing clinical trials, the bone marrow transplant register, and NHS databases. After the CED periods, NICE would undertake the re-assessment of the drug's cost-effectiveness [12].

5.7.2 CED and Cap on the Patient Number for Benlysta® (Belimumab) in Systemic Lupus Erythematosus

Uncertainty of clinical outcomes for Benlysta in systemic lupus erythematosus led UK payers to sign a CED scheme with GSK in 2016. Under the agreement, NHS would fund the drug during four years of data collection at a discounted price. Further, the deal assumed that the treatment would be continued beyond 24 weeks only if a pre-defined patient response was achieved. A cap on the number of treated patients was also agreed to contain budget impact [60].

5.7.3 Indication-Specific Pricing for Humira® (Adalimumab) in Hidradenitis Suppurativa

Abbvie signed an indication-specific pricing MEA with UK payers for Humira in 2016. The deal was driven by the uncertainty around the cost-effectiveness concluded in NICE's assessment of the drug in the hidradenitis suppurativa indication. Since the drug was also registered in other indications, the MEA assumed that a 20% discount would be applied only for this indication. A treatment stopping rule based on patient response was also written into the contract [61].

5.7.4 P4P and Discount for Mavenclad® (Cladribine) in Multiple Sclerosis

A P4P contract was signed by the UK payer with Merck for Mavenclad in 2017. The deal was conceived under the Accelerated

Access Review (AAR) passed by the government in 2016. Under the agreement, NHS England would only pay for the drug for patients who achieved a treatment response. Moreover, a discount was agreed with the manufacturer. Importantly, under the AAR, access to the treatment was granted immediately, without the regular three-month delay period [62].

5.7.5 P4P for Olysio® (Simeprevir) in Hepatitis C

Janssen-Cilag signed a P4P deal with the UK's payer in 2014. Under the deal, the company would refund the cost of Olysio for patients who did not clear the virus within 12 weeks and would sponsor a companion diagnostic test that could predict if the drug would be effective before treatment is started. Despite the success of the P4P, the drug was withdrawn from the market in 2018 due to the arrival of more effective direct-acting antiviral combinations [63].

CED and budget cap per patient (2017) for Strensiq® (asfotase alfa) in hypophosphatasia; CED and discount (2016) for Translarna® (ataluren) in Duchenne muscular dystrophy; CED, discount and budget cap (2015) for Vimizim® (elosulfase alfa) in mucopolysaccharidosis type IVA

CED schemes were signed as a part of NICE assessment of orphan drugs with limited clinical data. The deals were concluded for Strensiq in hypophosphatasia (2017), Translarna in Duchenne muscular dystrophy (2016) and Vimizim in mucopolysaccharidosis type IVA (2016). All CED deals shared common elements:

- five-year data collection period,
- defined treatment starting and stopping rules,
- planned re-assessment based on newly collected data,
- a financial agreement (discount, spending cap per patient).

These mixed-type MEAs enabled the management of both, financial and clinical uncertainties before a final reimbursement decision is reached [15, 63–65].

5.7.6 P4P for Maviret® (Glecaprevir/Pibrentasvir) and Sovaldi® (Sofosbuvir) in Hepatitis C

Due to their high cost, direct-acting antiviral drugs had been rationed by the NHS England to only selected patients, despite a positive NICE recommendation. To manage public pressure and improve patient access, NHS England signed a P4P deal with manufacturers in 2017. Under this MEA, NHS would be refunded by the manufacturers for patients who complete their treatment but do not achieve a cure (sustained virologic response). Patients were followed in a dedicated hepatitis C registry to monitor treatment uptake and outcomes and calculate paybacks from manufacturers [56].

5.7.7 CED for Spinraza® (Nusinersen) in Spinal Muscular Atrophy (SMA)

In 2019, NICE raised several concerns in the appraisal of Spinraza, concerning clinical data collection, PRO data collection and resource utilisation data collection. A CED deal was signed with the manufacturer for five years, with a minimum of three years' data collection. The final reimbursement decision was scheduled to take place by year 5 of the CED scheme. Several endpoints were collected for the CED study from a few sources: ongoing trials, the SMA REACH hospital registry, the NHS Blueteq System for high-cost drugs and patient-reported outcome measures being developed. Data was analysed twice per year according to a plan developed by the manufacturer. Data collection and analysis costs were borne by the manufacturer [13].

5.8 USA

US payers generally do not embrace MEAs as enthusiastically as their European counterparts. However, the US payer environment is highly fragmented and each of the private insurers may engage in confidential negotiations with manufacturers to manage their pharmacy budgets. They typically employ Pharmacy Benefits Managers who negotiate in-hospital medicine costs with companies. Those usually are concluded as simple confidential discounts. One of the first outcome-based MEAs was a 1998 P4P scheme for simvastatin, where Merck would refund patients and insurers the cost of the drug if it did not help them lower LDL cholesterol levels [66]. Other early P4P examples include the 2009 deals for sitagliptin and risedronate, where payments were linked to treatment outcomes in diabetes and osteoporosis correspondingly [67]. Commonly, the same P4P deal can be signed by the company with multiple private insurers [68].

Apart from private insurers, the government Centers for Medicare and Medicaid Services (CMS) is also increasingly facing financial pressure from novel costly treatments. While, legally, this public agency is not allowed to negotiate or control medicine prices, it found a way out of the growing pharmacy expenditure by implementing MEAs that were not straightforward price negotiations. Initially, CMS was chiefly engaged in CED deals in which CMS would cover a therapy only in the context of a clinical study. To date, CEDs have been implemented most often for non-drug technologies, such as diagnostics and medical devices [67]. In 1997, CMS signed its first P4P deal for Epoetin alfa, where reimbursement was linked to haemodialysis patients' outcomes [69]. Much later, in 2005, CMS signed its first CED contract for oncology drugs for the off-label treatment of colorectal cancer only as a part of clinical trials, to generate required evidence [67, 69]. However, since 2013 there has been a transition towards P4P deals, which became as common as CED [68]. Nevertheless, non-drug technologies continue to outnumber drugs among MEAs agreed by CMS. Also, there is a growing trend in CMS allowing state divisions to conclude their own outcome-based MEAs [70].

In terms of very costly orphan drugs, while the majority of products is reimbursed by private payers, CMS coverage can be patchy and patient co-payments are very high. Therefore, CMS is increasingly willing to explore new models for drug payment to enable their consistent coverage through Medicaid and in Medicare Parts B (ambulatory sector) and D (pharmacy benefit) [15]. This is particularly important for novel drugs with limited efficacy data but with significant budget impact.

Illustratively, in 2021, the US Food and Drug Administration (FDA) authorised Aduhelm (aducanumab) through the Accelerated Approval pathway, but with limited clinical efficacy data [71]. The approval was conditional on Biogen completing additional clinical trials. With questionable efficacy and a $56,000 annual price tag, the drug faced reimbursement hurdles, despite being the first Alzheimer's disease treatment on the market in nearly two decades. Given that 80% of the eligible patient population was covered by Medicare Part B, the annual spending of the public payer on Aduhelm was estimated to be over $100 billion. Whereas CMS is not allowed to incorporate cost-effectiveness in coverage decisions, regional managers may still be willing to restrict drug use to curb their budgets. To avoid such coverage restrictions, CMS will need to decide if a CED deal would be an appropriate option that can allow containing expenditure while ensuring broad patients access to the drug [72]. Such CED would take a form of a large randomised trial against placebo or the standard of care. However, given that US patients may be liable for co-pays for the novel drug even if it's reimbursed largely by Medicare as a part of the CED, it's problematic to envision that some patients incur co-pays for the new drug but receive placebo or standard care. Further, since the drug is indicated for minimally symptomatic patients, it may be difficult to demonstrate clinically meaningful benefit on cognition and function scales, as the study would be essentially a prevention trail.

5.8.1 P4P for Kymriah® (Tisagenlecleucel) in B-cell ALL

In 2017, Novartis signed a P4P agreement with CMS under which the payer will only pay for the childhood leukaemia therapy, Kymriah if patients respond to treatment within one month [15, 73]. A year after the agreement, CMS cancelled this contract without disclosing the reasons [70].

5.8.2 P4P for Entresto® (Sacubitril/ Valsartan) in Heart Failure

Between 2015 and 2017, Novartis signed a P4P deal with at least three insurers (Aetna, Cigna, Harvard PilgrimHealth Care) where drug payments were linked to the reduction in the proportion of patients with heart failure hospitalisations [68, 74].

5.8.3 P4P for Repatha® (Evolocumab) in Cardiac Arrest Prevention

Between 2015 and 2017, Amgen signed a P4P deal with at least four insurers (Cigna, CVS Health, Harvard Pilgrim Health Care, Prime Therapeutics) for the lipid-lowering drug Repatha. Under the deal, the company would refund in full the drug cost for each patient who had a heart attack or stroke while taking the drug. In addition, payers secured also a price discount [68, 75].

5.8.4 Population-Based P4P for Trulicity® (Dulaglutide) in Type 2 Diabetes

Whereas P4P typically involve rebates that are paid by the companies for individual, non-responding patients, this 2014 deal between Eli Lilly and Harvard Pilgrim assumed up or down price adjustments, depending on the Trulicity's performance in populations of diabetes patients. Under the terms of the deal, the insurer would pay a lower net price if fewer patients taking Trulicity achieved a specified HbA1c value compared with patients taking competitor treatments from the same therapeutic class, and a higher net price if patients taking Trulicity had better outcomes than patients taking other drugs [69].

5.8.5 P4P for Iressa® (Gefitinib) in Non-Small Cell Lung Cancer

Under this 2015 P4P deal, AstraZeneca would reimburse the pharmacy benefit management organization, Express Scripts, the costs of Iressa if a patient stopped treatment before the third prescription fill. Therefore, the payer avoided any pharmaceutical costs for patients who discontinued the treatment. Interestingly, in contrast to European P4P deals that often involve physicians in patients' data collection and rely on dedicated registries, treatment outcome measures for this deal were assessed indirectly, using claims data [69].

5.8.6 P4P for Enbrel® (Etanercept) in Rheumatoid Arthritis

In 2017, Harvard Pilgrim signed a two-year P4P contract with Amgen that involved six patient outcomes, such as patient compliance, switching or adding drugs, dose-escalation and steroid interventions. If patient scores on these outcomes were below a specified level, the insurer would receive a partial refund for the drug. The insurer sought to develop this deal because historically, only about a third of patients on Enbrel or similar drugs met all six criteria [76].

5.8.7 Preferential Formulary Status for Orbactiv® (Oritavancin) in Bacterial Skin Infections

Whereas P4P typically revolve around refunds for non-responding patients, this 2018 deal of Melinta Therapeutics with Oklahoma Medicaid assumed benefits to the manufacturer, conditional on the drug use not resulting in increased total health care costs (including hospitalisation) of treating patients for bacterial skin infections, compared to the current standard of care witch cheaper antibiotics. In exchange, Orbactiv received preferred status on the formulary and no longer required prior authorisation for use [73, 77].

5.8.8 PVA Linked to Patient's Adherence to Aristada® (Aripiprazole Lauroxil) in Schizophrenia

Whereas PVAs typically assume price reductions to payers that increase with the growing sales volume of a drug, this innovative MEA for Aristada, signed with Oklahoma Medicaid in 2018, linked additional rebates to payers for each prescription filed by a patient every two months. This one-year contract assumed that, as long as patients adhere to treatment, this result in greater drug sales. Consequently, greater sales volume would translate into decreases in the price the state pays for the drug. However, rather than linking the rebates to aggregated sales volume, this MEA linked price reductions to continuous sales that pertained to individual patients [70, 73].

5.9 CANADA

The majority of MEA contracts in Canada were concluded before the establishment of the pan-Canadian Pharmaceutical Alliance (pCPA) in August 2010. pCPA conducts joint provincial/

territorial/federal negotiations for branded and generic drugs to leverage the combined negotiating power of participating stakeholders. Several initiatives have emerged in the rare diseases area, e.g., an initiative by the Canadian Association for Population Therapeutics (CAPT) of the Expensive Drugs for Rare Diseases (EDRD) Working Group.

Payers in Canada prefer financial agreements to outcome-based ones, as they allow them to achieve cost containing with simpler means. This helps them avoid the complex organisational issues of funding, data collection and safety and the national scale-up of the MEAs. Individual drug discounts account for 95% of MEA contracts in Canada. Utilisation cap and PVAs are present in up to 30% and 15% of agreements, respectively. Since the establishment of the pCPA, 200–250 agreements were completed with only a handful of innovative contracts, such as indication-specific pricing, conditional treatment continuation, CEDs, and provisions that improve payer cash flow. There are a few innovative contracts at the national level that have proven to pose a significant administrative burden.

Canadian payers will have to adjust to ATMPs that are entering the market as current models are insufficient. However, it is unlikely that payers will make investments into infrastructure or resources to manage MEAs. Alberta, BC, Ontario has generally well-developed infrastructure and payers in this jurisdiction may consider clinical performance-based agreements. Further, while initially, the Patient Support Programs (PSPs) were relative to the financial aspects of drug funding, currently, there is more openness to use PSPs also for data collection.

In terms of successful implementations in Canada, MEAs are well suited for disease areas with small and well-defined patient populations (e.g., rare diseases, targeted therapies in oncology) where patients can be easily qualified for therapy and tracked. Further, these are treatment areas restricted funding would likely lead to public unrest and pressure to make them available. Drug performance measures and treatment start and stop criteria for MEA deals must be pre-specified. Also, MEAs should include a strong clinical endpoint and not a surrogate endpoint and should address cost containment and budget management issues. Strong support from the clinician community for the MEA is of great importance as well. Value-added contracts that improve the infrastructure or care management solutions may be also welcomed by Canadian payers.

In terms of hurdles to the implementation of outcome-based MEAs, the current system lacks infrastructure, resources and skillsets. Without investment, it will become even less capable of accommodating MEAs.

5.9.1 CED for Fabrazyme® and Replagal® (Agalsidase α and β) in Fabry Disease

In 2005, Canadian authorities entered into a CED agreement with Genzyme and Shire for enzyme replacement therapies in Fabry disease. This agreement was a means of addressing the public pressure after the initial negative reimbursement decision. Reimbursement was granted conditionally until new data on the treatments and disease were collected in a post-marketing study called Canadian Fabry Disease Initiative (CDF). This 10-years study formed the most complete prospective database on Fabry disease. CDFI outcomes data supported the final positive reimbursement decision [78–80].

5.9.2 Therapeutic Class Discounts for Hepatitis C Drugs

In 2017, the pan-Canadian Pharmaceutical Alliance signed a therapeutic class discount agreement for three hepatitis C drugs with Gilead, Merck, and Bristol-Myers Squibb. The deal secured lower prices for several hepatitis C drugs, including 4 new entities, than the provinces had negotiated individually. In some jurisdictions, the savings were used to cover treatment also for the less severe patients [81, 82].

5.9.3 P4P for Revestive® (Teduglutide) in Short Bowel Syndrome

To overcome a negative cost-effectiveness assessment of Revestive in short bowel syndrome by the CADTH, Shire signed a P4P agreement with the pan-Canadian Pharmaceutical Alliance. The scheme involved a patient support program, that assumed a payment only for responders. The drug was made available to patients at the beginning of 2019 [84, 85].

5.10 CONCLUSIONS

Overall, the most frequent motivation of payers to introduce MEAs are budget impact constraints, followed by uncertainty about clinical outcomes or real-life outcomes. Italy has the longest of experience in MEAs and more deals are foreseen in this country. While France, Sweden and UK have a moderate level of experience, they are increasingly disinclined towards outcome-based deals. Also, with significant experience in CED deals, Netherlands is expected to adopt more P4P schemes in the future. Canada and Germany have generally avoided MAEs and this situation is unlikely to change. Similarly, while Spain has a relatively good track record with outcome-based MEAs, mainly in the Catalonia region, the national scale-up of this trend is improbable. In the US, one has to consider separately the private insurers and the public payer CMS. While the former players have the liberty to sign a variety of financial deals that typically involve discounts and P4P, the latter one is more restricted by regulations. CMS has recently shifted from CED to more P4P deals but continues to prefer MEAs for non-drug technologies.

Overall, the experience of many countries indicates that CED agreements often fail to deliver their stated goal of reducing uncertainty around product performance [90]. Further, given their confidential nature, these deals are not the source of sufficient collaboration and information sharing between countries. This hampers the exchange of experiences that could lead to improved design of various performance-based schemes.

5.10.1 Key Success Factors for Innovative MEAs

The experience of healthcare payers with performance-based MEAs indicates that those should only be used where the benefit of additional evidence justifies the cost of executing MEAs [85]. Further, such MEAs must be designed so that they can address the uncertainties related to the products and lead to actionable outcomes, ensuring appropriate levels of transparency. Further, payers increasingly agree on MEAs that are a mix of various types of deals, that manage simultaneously the price, usage, payment structure, and conditions of use [86].

Companies and payers must keep in mind that the involvement of a strong, well-organised physician community is a very important ingredient for success. Outcome-based MEAs often involve extra resources and time use. This has to be balanced by a high unmet need or an important public health concern that is addressed by the medicine.

Also, the patient population in outcome-based MEAs needs to be easily identifiable and traceable. Therefore, medicines for rare diseases, targeted oncology therapies and ATMPs are suitable for such deals.

In terms of outcomes, surrogate endpoints are not suitable for MEAs. An established endpoint that demonstrates treatment outcomes within three to six months is necessary, such as cure or change in tumour size. For oncology areas with short life expectancy, survival may also be an appropriate endpoint.

As far as CED MEAs are concerned, they need to be able to address the underlying uncertainty around clinical or cost-effectiveness in a definitive way, to enable final pricing and reimbursement decisions. In this regard, simplicity is often valued by payers. Interestingly, a checklist has been proposed to assess if CED would be appropriate for a rare disease treatment [87].

However, given the large number of MEAs agreed across various geographies and the fact that the deal details are often spread over a large number of grey literature sources, the Artificial Intelligence (AI) is emerging as a promising tool in decision making. AI has been shown to accelerate and improve the quality of systematic literature reviews and also to extract the relevant information of large number of HTA documents published by national regulators. Further, AI tools are capable of predicting HTA decisions based on the analysis of historical appraisal documents, as well as proposing the correct comparators and position the therapy in the right place in the treatment pathway and identify the best patient profiles [88–90].

5.10.2 Key Hurdles for Innovative MEAs

The administrative burden of the deal for physicians and payers is a key factor to consider while contracting MEAs in France, Germany, Sweden, and the UK. Ideally, the extra efforts must be compensated by the generation of useful data that can be applied by healthcare administration (e.g., the size of the target population) or by the clinical community.

When dealing with the lack of experience of some payers, manufacturers need to show a proactive approach to negotiations and come up with ready solutions.

Some constraints are also due to financial and accounting rules. The majority of healthcare systems have a short-term budget horizon. Particularly, in Spain or Netherlands, only contracts that can be resolved within a fiscal are allowed.

Addressing the support of the clinical community is another key factor. In some countries, physicians may ask for extra time and compensation to manage outcome-based MEAs. This also should be addressed in negotiations.

The quality and timely delivery of patient registries for outcome-based MEAs is another key factor. While Italy and the Netherlands have significant experience with this regard, other countries may rely entirely on the manufacturer to establish and manage the registry. The company must engage with partners who have the necessary experience and skills to build such registries. Also, there have been concerns about the safety of collected data in Canada, France, Germany, and Sweden. Therefore, new regulations may be needed to resolve this issue.

Further, CED schemes are not immune to abuse from the side of the manufacturers. There have been cases of them intentionally delaying the start of data collection, data analysis protocols are sometimes not followed or have inherent design flaws, and the quality of collected data is sometimes unacceptable. Such bad experiences can effectively deter payers from entering into future MEAs.

REFERENCES

1. CEPS. Rapport d'activité 2016. 2017. Available from: http://solidarites-sante.gouv.fr/IMG/pdf/rapport_annuel_2016_medicaments.pdf
2. French Govt Must Act To Stop 'Inexorable Decline' of Pharmaceutical Industry: PinkSheet; 2018. Available from: https://pink.pharmaintelligence.informa.com/PS122346/French-Govt-Must-Act-To-Stop-Inexorable-Decline-Of-Pharmaceutical-Industry.
3. Garrison LP Jr., Towse A, Briggs A, de Pouvourville G, Grueger J, Mohr PE, et al. Performance-Based Risk-Sharing Arrangements-Good Practices for Design, Implementation, and Evaluation: Report of the ISPOR Good Practices for Performance-Based

Risk-Sharing Arrangements Task Force. Value Health: J. Int. Soc. Pharmacoecon. Outcomes Res. 2013;16(5):703–19.

4. Jaroslawski S, Toumi M. Market Access Agreements for Pharmaceuticals in Europe: Diversity of Approaches and Underlying Concepts. BMC Health Serv. Res. 2011;11:259.

5. Espín J, Rovira J, Garcia L. Experiences and Impact of European Risk-Sharing Schemes Focusing on Oncology Medicines. Eminet 2011. Available from: https://www.researchgate.net/profile/Jaime-Espin/publication/260343816_Experiences_and_Impact_of_European_Risk-Sharing_Schemes_Focusing_on_Oncology_Medicines/links/004635313806193b3e000000/Experiences-and-Impact-of-European-Risk-Sharing-Schemes-Focusing-on-Oncology-Medicines.pdf?origin=publication_detail

6. Fagnani F, Pham T, Claudepierre P, Berenbaum F, De Chalus T, Saadoun C, et al. Modeling of the Clinical and Economic Impact of a Risk-Sharing Agreement Supporting a Treat-to-Target Strategy in the Management of Patients With Rheumatoid Arthritis in France. J. Med. Econ. 2016;19(8):812–21.

7. Polyarthrite rhumatoïde: en cas d'échec d'un traitement avec Cimzia*, UCB rembourse l'assurance maladie: APM News; 2013. Available from: https://www.apmnews.com/nostory.php?uid=&objet=238792

8. Celgene a conclu un accord "efficace ou remboursé" avec le CEPS pour Imnovid* Paris: APM News; 2015. Available from: https://www.apmnews.com/Celgene-a-conclu-un-accord-efficace-ou-rembourse-avec-le-CEPS-pour-Imnovid-FS_256638.html

9. Dunlop WCN, Staufer A, Levy P, Edwards GJ. Innovative Pharmaceutical Pricing Agreements in Five European Markets: A Survey of Stakeholder Attitudes and Experience. Health Policy (Amsterdam, Netherlands) 2018;122(5):528–32.

10. Allen J. NOACs pricing, reimbursement, and access. London; 2017.

11. Comite economique des produits de santerapport d'activite 2012. Paris; 2013. Available from: https://solidarites-sante.gouv.fr/IMG/pdf/RA_2012_Final.pdf

12. Facey KM, Espin J, Kent E, Link A, Nicod E, O'Leary A, et al. Implementing Outcomes-Based Managed Entry Agreements for Rare Disease Treatments: Nusinersen and Tisagenlecleucel.

PharmacoEconomics 2021;39(9):1021–44. doi: 10.1007/s40273-021-01050-5. Epub 2021 Jul 7. PMID: 34231135; PMCID: PMC8260322

13. Jørgensen J, Hanna E, Kefalas P. Outcomes-Based Reimbursement for Gene Therapies in Practice: The Experience of Recently Launched CAR-T Cell Therapies in Major European Countries. J. Market Access Health Policy 2020;8(1):1715536.

14. IHS Markit report: Merck KGaA signs pay-for-performance contract with GKV funds for MS treatment Mavenclad, 2018.

15. Wesley T. Market Access Trends in Europe. Market Access by Region/Europe. Datamonitor Healthcare. Informa, 2018.

16. Jommi C, Addis A, Martini N, Nicod E, Pani M, Scopinaro A, Vogler S. Price and Reimbursement for Orphan Medicines and Managed Entry Agreements: Does Italy Need a Framework? Available from: https://journals.aboutscience.eu/index.php/grhta/article/view/2278

17. Navarria A, Drago V, Gozzo L, Longo L, Mansueto S, Pignataro G, et al. Do the Current Performance-Based Schemes in Italy Really Work? "Success Fee": A Novel Measure for Cost-Containment of Drug Expenditure. Value Health 2015;18(1):131–6.

18. Lucas F. Performance-Based Managed Entry Agreements for Medicines: Much Needed, but Not Feasible? Value Outcomes Spotlight 2016;Nov/Dec.

19. Garattini L, Curto A, van de Vooren K. Italian Risk-Sharing Agreements on Drugs: Are They Worthwhile? Eur. J. Health Econ.: Health Econ. Prevention Care 2015;16(1):1–3.

20. Gazzetta n. 289 del 12 dicembre 2012 Rome: AIFA; 2012. Available from: http://95.110.157.84/gazzettaufficiale.biz/atti/2012/20120289/12A12855.htm

21. Gazzetta n. 106 del 8 maggio 2012 Rome: AIFA; 2012. Available from: http://95.110.157.84/gazzettaufficiale.biz/atti/2012/20120106/12A04734.htm

22. Multi-indication Pricing: Pros, Cons and Applicability to the UK London: OHE; 2015. Available from: https://www.ohe.org/news/multi-indication-pricing-pros-cons-and-applicability-uk

23. Wilsdon T, Barron A. Managed entry agreements in the context of Medicines Adaptive Pathways to Patients London; 2016.

24. Lattanzi L, Avitabile A, Caprioli G, Caputo A, Stell A, Giuliani G. PCN323 – Perjeta Treatment Duration in the Italian Clinical Practice: A First Analysis of the Aifa Prescription Registry. Value Health 2018;21:S69.

25. Pani L, Cammarata S. The Italian Payers' approach to new anti-hepatitis C drugs, 2015.

26. Area Strategia ed Economia del Farmaco. 2017. Available from: http://www.aifa.gov.it/sites/default/files/Elenchi_farmaci_innovativi_fondi_Legge_Bilancio2017.pdf

27. Prada M, Mantovani M, Sansone C, Bertozzi C. Managed Entry Agreements for orphan drugs in Italy active on April 2016. European Conference on Rare Diseases & Orphan Products 26–28 May 2016 Edinburgh.

28. Gray N. GSK enters pay-for-performance deal for gene therapy Biopharma Dive; 2016. Available from: https://www.biopharmadive.com/news/gsk-enters-pay-for-performance-deal-for-gene-therapy/424650

29. Grubert N. Managed entry saved the Dutch healthcare system €588.3 million in 2020 LinkedIn: Personal blog; 2022. Available from: https://www.linkedin.com/feed/update/urn:li:activity:6900829398896123904/

30. Interview: Anita Atema – General Manager, Celgene, The Netherlands: PharmaBoardroom; 2016. Available from: https://pharmaboardroom.com/interviews/interview-anita-atema-general-manager-celgene-the-netherlands

31. Interview: Bart Vanhauwere – General Manager, Roche Netherlands: PharmaBoardroom; 2015. Available from: https://pharmaboardroom.com/interviews/interview-bart-vanhauwere-general-manager-roche-netherlands/

32. van den Brink R. Reimbursement of Orphan Drugs: The Pompe and Fabry Case in the Netherlands. Orphanet J. Rare Dis. 2014;9(1):O17.

33. Dutch "no cure no pay" scheme for some new drugs: PharmaTimes; 2012. Available from: http://www.pharmatimes.com/news/dutch_no_cure_no_pay_scheme_for_some_new_drugs_976705

34. Makady A, van Veelen A, de Boer A, Hillege H, Klungel OH, Goettsch W. Implementing Managed Entry Agreements in Practice: The Dutch Reality Check. Health Policy (Amsterdam, Netherlands). 2019;123(3):267–74.

35. CatSalut. Experiencias en acuerdos de riesgo compartido y esquemas de pago basados en resultados en Cataluña, 2015. Available from: http://catsalut.gencat.cat/web/.content/minisite/catsalut/proveidors_professionals/medicaments_farmacia/acords_risc_compartit/AES-2015_poster_68_ARC_def_09_06_2015.pdf

36. MSSSI. El Gobierno aprueba el primer tratamiento de Atrofia Muscular Espinal, cumpliendo su compromiso con los afectados por esta enfermedad, 2018. Available from: https://www.msssi.gob.es/en/gabinete/notasPrensa.do?id=4292

37. Diario Farma. Cruz inaugura, con Spinraza, una nueva era en financiación de fármacos, 2018. Available from: https://www.diariofarma.com/2018/02/06/cruz-inaugura-spinraza-una-nueva-financiacion-farmacos

38. Clopes A, Gasol M, Cajal R, Segú L, Crespo R, Mora R, et al. Financial Consequences of a Payment-by-Results Scheme in Catalonia: Gefitinib in Advanced EGFR-Mutation Positive non-Small-Cell Lung Cancer. J. Med. Econ. 2017;20(1):1–7.

39. Senior M. Managed entry agreements. London: Datamonitor Healthcare; 2014.

40. Rojas García P, Antoñanzas Villar F. Los Contratos De Riesgo Compartido En El Sistema Nacional De Salud: Percepciones De Los Profesionales Sanitarios. Revista Española de Salud Pública 2018;92.

41. Fontanilla D. Cimzia. London: Datamonitor Healthcare; 2018. Available from: https://pharmastore.informa.com/product/cimzia/

42. Grubert N. Key Trends in European Market Access. London; 2016. Available from: https://pharmastore.informa.com/product/datamonitor-key-trends-in-european-market-access/

43. Grubert N. Managed entry agreements (MEAs) provisionally reduced pharmaceutical expenditure in Sweden by SEK 2.7bn (€254mn) in 2021. LinkedIn: Personal blog; 2022. Available from: https://www.linkedin.com/feed/update/urn:li:activity:6901429624296075264/?msgControlName=reply_to_sender&msgConversationId=2-MDc4ZThmMzUtODBmNC-00Zjk3LWFhZjItMDJkMDEyZmU3NDc3XzAxMg%3D%3D&msgOverlay=true

44. Secretary General for Health Ministry of Health Directorate General for Basic NHS Services Portfolio and Pharmacy

Pharmacoclinical Protocol for the Use of Duplilumab in Severe Atopic Dermatitis in Adult Patients in the National Health System. Madrid; Spanish Ministry of Health, 2020. Available from: https://www.sanidad.gob.es/en/profesionales/farmacia/valtermed/docs/20200131_I_Protocolo_dupilumab_dermatitis_atopica__grave_adultos.pdf

45. Informe de Resultados de Los Pacientes Con Dermatitis Atopic Grave en Tratamiento Con Dupilumab Registrados en Valtermed. Madrid; Spanish Ministry of Health. February 2022. Availabe from: https://www.sanidad.gob.es/profesionales/farmacia/valtermed/docs/20220228_Informe_Valtermed_Dupilumab.pdf

46. Besparingar från sidoöverenskommelser 2021 Slutavstämning, TLV, Stockholm, March 2022. Available from: https://www.tlv.se/om-oss/press/nyheter/arkiv/2022-03-18-slutavstamning-av-besparingar-fran-sidooverenskommelser-helaret-2021.html

47. Willis M, Persson U, Zoellner Y, Gradl B. Reducing Uncertainty in Value-Based Pricing Using Evidence Development Agreements: the Case of Continuous Intraduodenal Infusion of levodopa/carbidopa (Duodopa®) in Sweden. Appl. Health Econ. Health Policy 2010;8(6):377–86.

48. The development of pharmaceutical expenditure in Sweden. Stockholm: TLV; 2017. Available from: https://www.tlv.se/download/18.6919e22e161936e05891e287/1518688820277/development_of_pharmaceutical_expenditure_in_sweden.pdf

49. TLV assessment report. 2017. Available from: https://www.tlv.se/beslut/beslut-lakemedel/begransad-subvention/arkiv/2017-09-29-maviret-ingar-i-hogkostnadsskyddet-med-begransning.html

50. Medivir concludes Swedish agreement on Olysio-based treatment for hepatitis C Pharma Letter; 2014. Available from: https://www.thepharmaletter.com/article/medivir-concludes-swedish-agreement-on-olysio-based-treatment-for-hepatitis-c

51. Vertex Announces Long-Term Access Agreement in Sweden for Cystic Fibrosis Medicine ORKAMBI® (lumacaftor/ivacaftor): Business Wire; 2018. Available from: https://www.businesswire.com/news/home/20180618005495/en/Vertex-Announces-Long-Term-Access-Agreement-Sweden-Cystic

52. Jaroslawski S, Toumi M. Design of Patient Access Schemes in the UK: Influence of Health Technology Assessment by

the National Institute for Health and Clinical Excellence. Appl. Health Econ. Health Policy 2011;9(4):209–15.

53. Bovenberg J, Penton H, Buyukkaramikli N. 10 Years of End-of-Life Criteria in the United Kingdom. Value Health 2021;24(5):691–8.

54. Pharmaphorum. Ultra-rare disease drugs: has access in England just got harder? 2017. Available from: https://pharmaphorum. com/views-and-analysis/ultra-rare-diseases-england/

55. Gornall J, Hoey A, Ozieranski P. A Pill Too Hard to Swallow: How the NHS Is Limiting Access to High Priced Drugs. BMJ. 2016;354:i4117.

56. 25,000 Hepatitis C patients receive new treatments: NHS England; 2018. Available from: https://www.england.nhs.uk/ blog/25000-hepatitis-c-patients-receive-new-treatments/

57. Longworth L, Youn J, Bojke L, Palmer S, Griffin S, Spackman E, et al. When Does NICE Recommend the Use of Health Technologies Within a Programme of Evidence Development?: A Systematic Review of NICE Guidance. PharmacoEconomics 2013;31(2):137–49.

58. Walker S, Sculpher M, Claxton K, Palmer S. Coverage With Evidence Development, Only in Research, Risk Sharing, or Patient Access Scheme? A Framework for Coverage Decisions. Value Health 2012;15(3):570–9.

59. Ibrance v Kisqali: Quicker Novartis Discount Helps Cut UK NICE's Timeline: In Vivo. Pharma Intelligence; 2017. Available from: https://invivo.pharmaintelligence.informa.com/PS121987/ Ibrance-v-Kisqali-Quicker-Novartis-Discount-Helps-Cut-UK-NICEs-Timeline

60. Benlysta, Managed Access Agreement London: NICE; 2016. Available from: https://www.nice.org.uk/guidance/ta397/resources/ managed-access-agreement-september-2016-pdf-2665741069

61. Tappenden P, Carroll C, Stevens JW, Rawdin A, Grimm S, Clowes M, et al. Adalimumab for Treating Moderate-to-Severe Hidradenitis Suppurativa: An Evidence Review Group Perspective of a NICE Single Technology Appraisal. Pharmacoeconomics 2017;35(8):805–15.

62. NHS England partners with Merck on a commercial agreement that allows people with MS in England immediate access to cladribine tablets (Mavenclad®): ACNR; 2017. Available

from: http://www.acnr.co.uk/2017/11/nhs-england-partners-with-merck-on-a-commercial-agreement-that-allows-people-with-ms-in-england-immediate-access-to-cladribine-tablets-mavenclad/

63. Kusel J, Spoors J. Recent Trends in the Pricing of High-Cost Pharmaceuticals. Br. J. Healthcare Manag. 2016;22(5):267–77.

64. Spoors J, Kusel J. The Evolution of Patient Access Schemes. Value Health 2016;19(7):A462.

65. Ismailoglu I, Duttagupta S. Divergence of Evaluation of Orphan Drugs Between Regulators and Payers: Implications For Patient Access In US And EU. Value Health 2017;20(9): A571–A2.

66. Møldrup C. No Cure, No Pay. BMJ (Clinical research ed). 2005;330(7502):1262–4.

67. Carlson JJ, Sullivan SD, Garrison LP, Neumann PJ, Veenstra DL. Linking Payment to Health Outcomes: A Taxonomy and Examination of Performance-Based Reimbursement Schemes between Healthcare Payers and Manufacturers. Health Policy (Amsterdam, Netherlands). 2010;96(3):179–90.

68. Carlson JJ, Chen S, Garrison LP Jr. Performance-Based Risk-Sharing Arrangements: An Updated International Review. Pharmacoeconomics 2017;35(10):1063–72.

69. Yu JS, Chin L, Oh J, Farias J. Performance-Based Risk-Sharing Arrangements for Pharmaceutical Products in the United States: A Systematic Review. J. Manag. Care Specialty Pharm. 2017;23(10):1028–40.

70. Oklahoma Signs the Nation's First State Medicaid Value-Based Contracts for Rx Drugs Portland: The National Academy for State Health Policy; 2018. Available from: https://www.nashp.org/oklahoma-signs-first-medicaid-value-based-contracts-for-rx-drugs/

71. Cohen J. Controversial FDA Approval of Alzheimer's Drug Aducanumab Sets Stage For Possible Medicare Coverage Battle: Forbes; 2021.

72. Medicare 'Coverage With Evidence Development' For Aducanumab? How Might It Work?: Health Affairs Blog; 2021. June 30, 2021. Available from: https://www.healthaffairs.org/do/10.1377/hblog20210625.284997/full/

73. Seeley E, Chimonas S, Kesselheim AS. Can Outcomes-Based Pharmaceutical Contracts Reduce Drug Prices in the US? A Mixed Methods Assessment. J. Law Med. Ethics 2018;46(4):952–63.

74. Cigna Implements Value-Based Contract with Novartis for Heart Drug Entresto: Cigna; 2016. Available from: https://www.cigna.com/about-us/newsroom/news-and-views/press-releases/2016/cigna-implements-value-based-contract-with-novartis-for-heart-drug-entrestotm

75. Harvard Pilgrim strikes 'pay-for-performance' deal for cholesterol drug: Boston Globe; 2015. Available from: https://www.bostonglobe.com/business/2015/11/08/harvard-pilgrim-strikes-pay-for-performance-deal-for-cholesterol-drug/iGIV7rBie4K20HNbKORsPJ/story.html

76. Harvard Pilgrim Signs Outcomes-Based Contract with Amgen for Enbrel: Harvard Pilgrim; 2017. Available from: https://www.harvardpilgrim.org/public/news-detail?nt=HPH_News_C&nid=1471912468296

77. Oritavancin outcomes based contract with Oklahoma Medicaid may be the new normal: Contagion Live; 2018. Available from: https://www.contagionlive.com/view/oritavancin-outcomesbased-contract-with-oklahoma-medicaid-may-be-the-new-normal

78. Silversides A. Fabry Trial Set to Answer "political Problem". CMAJ 2009;181(6–7):365–6.

79. Silversides A. Enzyme Therapy for Fabry Patients in Jeopardy. CMAJ 2009;181(6–7):E120–E.

80. Bishop D, Lexchin J. Politics and Its Intersection with Coverage with Evidence Development: A Qualitative Analysis from Expert Interviews. BMC Health Serv. Res. 2013;13(1):88.

81. New Collective Drug Price Will Provide More Canadians Access to Hepatitis C Treatment Dallas, TX: Hepatitis News Today; 2017. Available from: https://hepatitisnewstoday.com/2017/03/02/new-collective-drug-prices-provide-more-canadians-access-hepatitis-c-treatment

82. A Statement from the pan-Canadian Pharmaceutical Alliance: Newswire; 2017. Available from: https://www.newswire.ca/news-releases/a-statement-from-the-pan-canadian-pharmaceutical-alliance-614373463.html

83. Cadth Canadian Drug Expert Committee Final Recommendation. Teduglutide (Revestive – Shire Pharma Canada ULC/NPS Pharma Holdings Ltd.) Indication: For the treatment of adult patients with Short Bowel Syndrome (SBS) who are dependent on parenteral support: CADTH; 2016. Available from: https://www.cadth.ca/sites/default/files/cdr/complete/SR0459_Revestive_complete_Jul-29_16.pdf

84. pCPA Monthly Trends & Insights – August 31, 2018 Toronto: MORSE Consulting; 2018. Available from: https://morsecon-sulting.ca/pcpa-monthly-trends-insights-august-31-2018/

85. Wenzl M, Chapman S. Performance-based managed entry agreements for new medicines in OECD countries and EU member states. 2019.

86. Vreman RA, Broekhoff TF, Leufkens HG, Mantel-Teeuwisse AK, Goettsch WG. Application of Managed Entry Agreements for Innovative Therapies in Different Settings and Combinations: A Feasibility Analysis. Int J Environ Res Public Health. 2020;17(22):8309.

87. Facey K. Checklist for a Rare Disease Treatment—Is an Outcomes-Based Managed Entry Agreement Feasible? Zenodo; 2021. Available from: https://doi.org/10.5281/zenodo.5032840

88. *Axess4you—inspired by patients, powered by artificial intelligence.* (n.d.).Axess4you. Accessed on January 11, 2022. Available from: https://www.axess4you.com/

89. Sakata Y, Inoue K, Nagasawa T, Ooishi M, Azuma M, Kitabayashi H, et al. PCN267 Further Development of Artificial Intelligence Supporting Systematic Literature Review for Conducting Cost-Effectiveness Analysis. Value Health. 2020;23:S470, Accessed on December 9, 2021.

90. Bonakdari H, Pelletier J-P, Martel-Pelletier J. A Reliable Time-Series Method for Predicting Arthritic Disease Outcomes: New Step from Regression toward a Nonlinear Artificial Intelligence Method. Comput. Methods Programs Biomed. 2020;189:105315.

Chapter 6

Novel Funding Models for Expensive Therapies

The arrival of novel, innovative health technologies, such as gene and cell therapies, requires specific funding solutions. Whereas such products come at unprecedentedly high cost, they have the potential to cure patients and provide lifelong benefits with a single or short-term administration.

While some of the benefits of Expensive Therapies (ETs) are only potential and not proven in long-term clinical trials, payers maintain that prices of some of these products are not justified. For example, some drugs are merely extensions of older products providing marginal benefits, and also, many patients receiving costly therapeutics do not see the expected health benefits [1].

In the US, novel financial models have been proposed to address this issue, such as amortisation, reinsurance, and rewarding therapy adherence with reduced co-payments [1]. They have the potential to reduce high upfront costs and to distribute them across various players of the healthcare system. Since in the US patients change health insurance providers often, payers may feel that they do not benefit from the long-term savings once and insured patients leaves to a different provider [2]. Some authors proposed that, if a therapy is capable of extending patients' life or improve their quality of life beyond the treatment duration, then payment schemes for

DOI: 10.1201/9781003048565-6

such therapies should reflect the extended benefits. However, these novel models still require proof-of-concept testing with a well-defined population. Below, we discuss various novel funding models for ET that have been proposed in literature.

6.1 AMORTISATION

Amortisation is an accounting technique that allows an intangible asset to be depreciated on the balance sheet and the value to be allocated over a predetermined number of consecutive years according to the amortisation schedule [3]. Amortisation may be applied to the cost of ET if they are treated as an intangible asset. At the same time, this technique allows ET manufacturers to be fully compensated for the therapy at the time of purchase. Amendments to the generally accepted accounting principles (GAAP) and International Financial Reporting Standards (IFRS) will be required for this approach to become applicable in the pharmaceutical market [3].

Alternatively, payer payments to manufacturers for ET could be spread out over time on an agreed basis or at specific milestones [3]. These payment milestones may be tied directly to patient outcomes, as they are followed over time. If at any specified time point, outcomes or agreed milestones are not reached, payments to the manufacturer may cease.

Other mentions of amortisation in the literature consider conditions on granting market exclusivity for a product, when the costs of research and development (R&D) are amortised through the use of the active ingredient of the pharmaceutical in treatment of another indication [4]. This could lead to lower prices for the drug, if it is approved in novel indications. Also, an economic modelling study proposed that amortisation can be applied to mitigate high upfront drug costs by transferring financial responsibilities among multiple payers over time [2].

6.2 BUNDLE PAYMENT AND EPISODE OF CARE

A bundle payment is an integrated single payment and reward that covers a defined scope of services, regardless of the quantity of care [5]. It involves payment for all services related to a specific treatment, condition or patient, either by procedure or episode of care. Under this payment system, healthcare providers (HCP) are incentivised to control drug expenditure, which may cause unjustified avoidance of the use of advanced ET. Therefore, these models need to incorporate quality metrics that ensure good quality of care. HCP can also team up to deliver comprehensive care in a specified disease area for patients. In the US, such integrated systems were incentivised by the Patient Protection and Affordable Care Act under the term Affordable Care Organizations (ACO). Further, the Oncology Care Model (OCM) was developed by Centers for Medicare & Medicaid Services (CMS) that combines the payments for the episode of oncology care with per-beneficiary per-month (PBPM) fee for each episode of chemotherapy, and a performance-based payment related to quality of care metrics [5]. The development of bundle payment for dialysis has led to a reduction of use/misuse of erythropoietin stimulating agent and improved overall patients care [6].

6.3 ANNUITY/INSTALMENT PAYMENTS

Annuity or instalment payments allow payers to pay the cost of an ET based on an instalment schedule, which may be annual or on some other schedule [7, 8]. Although annuity is the most used name, it is inappropriate as it does not have term and is a lifetime payment. In contrast, instalment payment has a fixed term and is therefore more appropriate. Payment can also be contingent on certain criteria, such as patient outcomes, under performance-based agreements. Under this model, payers ensure consistent and/or constant adequate cash flow, but still record on their books the high upfront costs in the year the therapy is purchased and administered. However, manufacturers' revenue is recorded on their books in the year of the sale, while the cash flow is spread over the term of the payment schedule.

This model is particularly interesting for therapies that are possibly curative, but lack long-term treatment outcome data. As such, payers are uncertain if the high upfront cost will be justified by long-term outcomes that are only potential. Therefore, they may seek to agree with manufacturers on long-term instalment payment schedules that correspond to the outcomes horizon.

6.4 HEALTHCOIN

As a new tradable currency, Healthcoin has been proposed to convert incremental outcomes achieved by an ET into a common currency, such as life-year equivalents [9]. This solution appears attractive for a system with multiple payers and a high turnover of insurance providers, such as in the US [3]. If a payer pays for an ET and the patient changes insurers, int's the next, not the previous insurer who will reap the long-term health benefits from the treatment. To solve this issue, the second or subsequent insurer would pay to the first plan for the transferred patient in Healthcoins. The amount paid would reflect the size of the incremental outcomes generated for the patient. Healthcoins would be exchangeable to the national currency on the Healthcoin market.

This system could be feasible if clear terms are established and adhered to by all payers participating in a Healthcoin market, which may include public and private insurers alike. Also, there may be inherent uncertainty about the long-term efficacy of the treatment or the permanent nature of the cure and the model does not take into account that the value of the Healthcoin may depend on the age of the patient.

Further, it is unclear whether the current legal framework in several countries would allow the introduction of the Healthcoin. For payers, this payment model strategy will not improve cash flow or the high upfront budget costs. However, it does encourage them to invest in novel ET with long-term benefits. There is no major direct impact to the ET manufacturer from the implementation of this payment model.

6.5 OUTCOME-BASED ARRANGEMENTS

These agreements, which are also discussed in a separate chapter, have been widely implemented in healthcare for various types of therapies, including innovative and high-cost therapies [3]. Essentially, these agreements between manufacturers and payers allow therapies to enter the market under certain pre-determined conditions that are directly tied to the outcomes these therapies are intended to deliver to patients. They may also be linked to an instalment payment. Nevertheless, there are several challenges in implementing outcome-based arrangements. In particular, for ETs that offer medium- to long-term benefits, outcomes must be clearly defined before signing the agreement, and the timeline for tracking outcomes can be much longer than for traditional, conventional therapies.

Under these agreements, payers may feel reassured that they are not paying for those who do not benefit from the therapy. In reality, this payment model can offset cash flow and uncertainty about effectiveness to some degree by linking payment for therapies to certain conditions. However, this payment model does not address the high upfront costs, because any possible rebates from manufacturers for non-responding patients are postponed until patients' outcomes are assessed. The impact on manufacturers could be significant under this payment model, depending on the performance of the ET. As long as the therapy yields the intended benefits, manufacturers are compensated for the therapy. If the therapy does not produce the intended results, depending on the terms of the agreement, manufacturers may be required to reimburse payers for the intended benefits not provided.

A particular consideration is warranted for the so-called coverage with evidence development (CED) [10]. Under this model, ET is provided to patients conditionally for a specified period, until new clinical evidence is produced from clinical trials or from observational studies. An additional escrow agreement between the payer and the manufacturer assumes that the drug payments are deposited in a bank account of a public institution until the CED results are available [5]. If they results prove drug's value, they are transferred in full to the manufactures. In the opposite case, only a pre-specified fraction of the money is released to the company and the rest is returned to the payer. Certainly, such schemes are only feasible if the novel data can be produced in a reasonably short time and is not appropriate from therapies that are supposed to generate life-long benefits to patients.

6.6 PATIENT LOAN FROM PRIVATE OR GOVERNMENT ORGANISATIONS

Under this payment model, patients are responsible for taking out a loan, sometimes referred to as a health care loan (HCL), making it easier for patients to access the costly therapy [3]. However, even if a patient can access such loan, it will have a long-term detrimental impact on their budget [5]. Also, loan repayments will cease immediately upon the patient's premature death. This means that if a patient dies, the balance of the loan will remain unpaid. Also, even if the patient does not receive the intended benefit from the therapy, they would still be obligated to repay the loan. This model is obviously outside the scope of a national or private health insurance.

6.7 PAYER LOAN FROM PRIVATE OR GOVERNMENT ORGANISATIONS

Similar to patient loans, payers may also obtain loans to finance costly therapies. Payers would be expected to repay these loans [3]. They could obtain the loans through various lending mechanisms, including from the government. These loans could also be repaid, provided the ETs are deemed repayable. In a single-payer system, the payer could already be the government, whose budget has already been set. However, it is unclear what would happen in a multi-payer system, such as in the US, and whether responsibility for payments and loans would transfer to the patient's new provider.

6.8 SPECIAL DEDICATED GOVERNMENTAL FUND

Under this payment model, a government may decide to establish special silo funds to finance ET [3]. Such funds already exist, such as the UK's Cancer Drug Fund, the Scottish "new medicines fund," and the Australian Complex Authority Required Highly Specialized Drugs Program. Such funds are usually established in single-payer systems, with their budget set in addition to and separate from the overall health insurance budget. These types of funds are usually financed from the state budget (e.g., through taxes) as they constitute only a small fraction of the national budget. Interestingly, in Italy, a "5% AIFA fund" is collected from the pharmaceutical industry and enables the financing of orphan ET.

However, in a fragmented system of payers, as in the US, it could prove more complex to implement. It might be feasible for the US Centers of Medicare & Medicaid Services to pass legislation that would allow for the establishment of such a fund. However, passing such a law could prove difficult due to opposition to increasing government intervention in health care.

Importantly, HCP and insurers are unlikely to risk such a high level of investment for innovative drugs that lack solid outcomes data or when there is uncertainty about long-term outcomes [5, 11].

Manufacturers clearly benefit from this model, as they are paid immediately and in full at the time the payer purchases the therapeutic.

6.9 INSURANCE POOL

Under this model, multiple health insurers in an area contribute jointly to a fund to finance expensive ET [3]. Such insurance may apply to all-private or public-private partnerships that aim to recover the costs of goods within a given time. In this way, the risk of disproportional distribution of patients who require costly therapies is mitigated. The size of the financial input may depend on the number of people covered and on risk factors, such as age. Insurance pooling has been implemented in Germany. Also, the so-called risk corridors have been proposed by the US Department of Health and Human Services [12]. In this model, the public entity collects funds from private insurance plans with lower-than-expected claims, and transfers the funds to plans with higher claims.

6.10 REINSURANCE

Reinsurance occurs when payers secure coverage in the event of large, unforeseen pay-outs for ET [3]. In this model, multiple payers purchase an insurance policy to protect against excess financial risk. Reinsurance payments can be made annually or on an agreed schedule. This option deals with the risk of disproportionate distribution of patients who need ET among multiple insurers [5]. This would be of interest if the number of ET and the number of covered patients become very large, because in these circumstances, the risk of disproportionate distribution of ETs is higher. Manufacturers would not be affected by this payment model. It could facilitate access to patients if payers are less concerned about the risk, because they are insured. Such a pool has been established in Germany for regenerative therapies [13].

6.11 INTELLECTUAL-PROPERTY BUY-OUT

Under this model, the payer buys the intellectual property (IP) rights to the ET from the manufacturer or becomes the licensee, in order to manage production and distribution of the ET [5, 11, 14]. In this way, the payer incurs high upfront costs of the IP, but avoids the problem of high drug prices. Nevertheless, the approach involves the risks related to the manufacturing and distribution of the ET and also, does not address the uncertainty related to the efficacy of the drug [15].

6.12 DISCUSSION

The most straightforward way of financing ET appears to be via the special dedicated governmental fund. Such funds represent separate money pool that exists in addition to the regular pharmaceutical budget. Drug inclusion criteria are also different from the traditional reimbursement path and are restricted to innovative ET that addresses an important medical need. The development of a dedicated value assessment framework may be needed to assess the value of ET. Such a special fund could also incorporate outcome-based payment models, discounts, rebates, price–volume agreements, cost sharing, and annuity payments to improve its sustainability. Among the outcome-based models, CED with escrow agreement should be preferred for ET approved with immature clinical data.

Funding sources for such funds are usually from general taxation, but the Italian fund is also alimented by contributions from the pharmaceutical industry. However, there also needs to be rules as to what percentage of GDP the country is willing to allocate to such fund.

It appears that the outcome-based payment models, along with annuity/instalment payments, as well as healthcare loans for payers offer a reasonable balance of feasibility and financial attractiveness to payers as well. Figure 6.1 explains the impact of instalment payment and amortisation on the health care payer's budget impact and on the cash flows of the payer and the drug manufacturers. As illustrated, a combination of amortisation and instalment payment would mitigate the impact of costly, potentially curative, single-administration medicines on both the budget impact and the cash flow of the health care payers. Various approaches to outcome-based strategies and annuity payments have been widely implemented across the world in the past two decades and they are discussed in a separate chapter. While certain types of these agreements turned out to be burdensome and bring about little cost savings, payers in the majority of countries have learnt their lessons and continue to implement such programs.

Amortisation is likely a useful way forward as it addresses budget impact for payers. However, it will require that the generally accepted accounting principles are revised to acknowledge that some therapies meet the predefined criteria as intangible assets. Such criteria need to be defined, and duration of amortisation will also need to obey specific model, yet to be developed.

However, annuity payments and payer credits may prove to be problematic as well. They do not address the issue of affordability but only shift payers' financial burden to the future, which will eventually challenge budget's sustainability. However, if payers' future revenues are expected to increase significantly, it may be a worthwhile solution to payers.

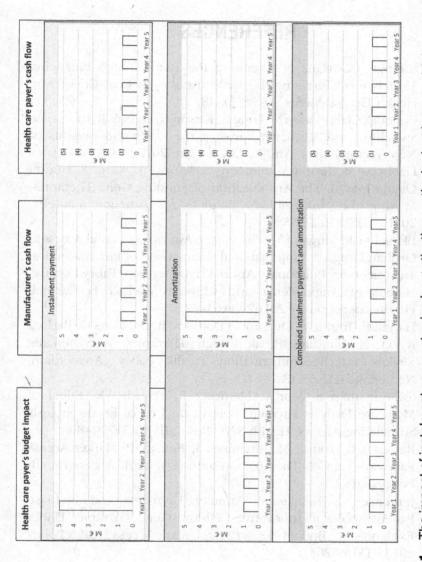

Figure 6.1 The impact of instalment payment or/and amortisation on the budget impact and the cash flows of a 5 million euro-cost product with a single administration, over a five-year time horizon.

On the contrary, patients' loans, Healthcoin, and IP buy-out strategies seem far more burdensome to implement and represent relatively little financial advantage to payers.

REFERENCES

1. Kleinke J, McGee N. Breaking the Bank: Three Financing Models for Addressing the Drug Innovation Cost Crisis. Am. Health Drug Benefits 2015;8(3):118.
2. Cutler D, Ciarametaro M, Long G, Kirson N, Dubois R. Insurance Switching and Mismatch between the Costs and Benefits of New Technologies. Am. J. Manag. Care 2017;23(12):750–7.
3. Dabbous M, Toumi M, Simoens S, Wasem J, Wang Y, Huerta Osuna J, et al. The Amortization of Funding Gene Therapies: Making the "Intangibles". Tangible for Patients. medRxiv. 2021:2021.04.16.21255597.
4. Blankart CR, Stargardt T, Schreyögg J. Availability of and Access to Orphan Drugs: An International Comparison of Pharmaceutical Treatments for Pulmonary Arterial Hypertension, Fabry Disease, Hereditary Angioedema and Chronic Myeloid Leukaemia. Pharmacoeconomics 2011;29(1):63–82.
5. Hanna E, Toumi M, Dussart C, Borissov B, Dabbous O, Badora K, et al. Funding Breakthrough Therapies: A Systematic Review and Recommendation. Health Policy (Amsterdam, Netherlands). 2018;122(3):217–29.
6. Swaminathan S, Mor V, Mehrotra R, Trivedi AN. Effect of Medicare Dialysis Payment Reform on Use of Erythropoiesis Stimulating Agents. Health Serv. Res. 2015;50(3):790–808.
7. Rémuzat C, Toumi M, Jørgensen J, Kefalas P. Market Access Pathways for Cell Therapies in France. J. Mark Access Health Policy 2015;3.
8. Jørgensen J, Kefalas P. Annuity Payments can Increase Patient Access to Innovative Cell and Gene Therapies Under England's Net Budget Impact Test. J. Mark Access Health Policy 2017;5(1):1355203.
9. Basu A, Subedi P, Kamal-Bahl S. Financing a Cure for Diabetes in a Multipayer Environment. Value Health: J. Int. Soc. Pharmacoecon. Outcomes Res. 2016;19(6):861–8.

10. Jaroslawski S, Toumi M. Market Access Agreements for Pharmaceuticals in Europe: Diversity of Approaches and Underlying Concepts. BMC Health Serv. Res. 2011;11:259.
11. Carr DR, Bradshaw SE. Gene Therapies: The Challenge of Super-High-Cost Treatments and How to Pay for Them. Regenerative Med. 2016;11(4):381–93.
12. Zettler PJ, Fuse Brown EC. The Challenge of Paying for Cost-Effective Cures. Am. J. Manag. Care 2017;23(1):62–4.
13. Hanna E, Toumi M. Gene and cell therapies: Market access and funding. 2020. Boca Raton, FL: CRC Press.
14. Jaroslawski S, Toumi M. Non-Profit Drug Research and Development: The Case Study of Genethon. J. Mark Access Health Policy 2019;7(1):1545514.
15. Jaroslawski S, Toumi M, Auquier P, Dussart C. Non-Profit Drug Research and Development at a Crossroads. Pharm Res. 2018;35(3):52.

Chapter 7

Managed Entry Agreement for Cell and Gene Therapies

Tingting Qiu and Mondher Toumi

7.1 EVOLVING DRUG DEVELOPMENT AND CHALLENGES IN CLINICAL TRIALS

To facilitate the market access of innovative products with high potential to fulfil unmet medical needs, more flexibility in accepting the evidence from the pivotal studies has been allowed by the regulators [1]. This is reflected in the ever increasing number of products that are granted with "conditional market authorization" in Europe, but also marketing authorisation under exceptional circumstances and adaptive licensing. According to the European Medicines Agency's (EMA) 10-year report on the conditional marketing authorisation (2006–2016) [2], phase I and/or phase II studies represent 55% of studies that were submitted as pivotal studies, and 34% of these studies are single-arm studies. There is also a trend that an increasing number of drugs were launched based on the studies with small patient numbers, which is in accordance with the increasing number of drugs with the orphan drug designation in the past two decades [3]. Additionally, more complexity is added

DOI: 10.1201/9781003048565-7

due to the introduction of innovative study design, such as the adaptative designs, which is referred to as relatively flexible clinical trial design including a pre-planned opportunity for modification of one or more specified aspects of the trial (e.g., sample size, randomisation ratio, number of treatment arms, the dose administered, etc.) usually based on interim data analysis or Bayesian statistics [4]. As a result, early market introduction with less comprehensive evidence or immature data has increased the uncertainty for payers/Health Technology Assessment (HTA) bodies to evaluate the product added value and conduct budget impact assessments. While regulators only assess the direction of the benefit-risk ratio (positive/negative), HTA and payers need to assess the magnitude of the additional benefit to appreciate the added value and related premium price.

7.2 LIMITATIONS IN THE CLINICAL EVIDENCE FOR CELL AND GENE THERAPIES

Cell and gene therapies, which are often perceived as breakthrough therapies that could bring promising treatment options for previously untreatable diseases, showed how regulators are open and willing to accept a higher level of evidence uncertainties. When investigating the pivotal studies used to support the market approval of 11 cell and gene therapies in Europe by EMA, single-arm studies with limited patient numbers and short follow-up durations, instead of well-designed double-blind randomised clinical trials (RCTs), constituted the pivotal evidence for seven products to gain market authorisation [1]. Moreover, patient-relevant endpoint (e.g., survival or morbidity) was reported as primary outcomes in only three products, with the remaining studies using surrogate endpoints as primary outcomes. The limitations in the clinical trials of cell and gene therapies increase the challenges in the reliable HTA clinical and economic analysis of cell and gene therapies. The length of follow-up studies was too short to evaluate the long-term data on survival, morbidity, and quality of life (QoL) [5]. The paucity of data from clinical studies raises difficulties in calculating the transition probabilities for feeding economic models, while very few data are available for health utility relevant health states. The lack of comparative evidence determines that assumptions must be made on the efficacy and effectiveness of cell and gene therapies against the current standard of care or usual care [6].

7.3 THE HIGH PRICE OF CELL AND GENE THERAPIES

Due to the complexity in the manufacturing, research, and development of cell and gene therapies, the high expectation of long-term benefits, and the potential for cure, they are generally associated with a higher price tag than conventional drugs to recoup the cost. The cell and gene therapies approved in Europe were priced at a range of $21,926–$2,125,000. Three gene therapies, Zolgensma ($2,125,000), Zynteglo ($1,800,000), and Luxturna ($850,000) were listed as the top three most expensive drugs in the US. Compared to conventional medicines, one unique feature of cell and gene therapies is that they will cause substantial upfront costs through a single or a limited number of administrations [7], making it not fitting with existing payment models designed for small molecules or biologics administered regularly [8]. The greater financial risk associated with cell and gene therapies is that there is no opportunity to discontinue treatment if they are proven to be ineffective or unsafe in the real world setting [9].

It is noteworthy that the drug price may account for only a small portion of the total costs of treatment with cell and gene therapies, which include also the additional administrative burdens associated with providing the therapies [10]. In the case of CAR-T cell therapies, in addition to the immediate acquisition price, other costs, including the cost for continuing infusions of intravenous immunoglobulin (IVIG) for hypogammaglobulinemia [11] and anti-IL-6-receptor antibody (e.g., tocilizumab) for cytokine release syndrome could not be marginal [10]. On top of drug costs, the logistic costs may be significant. Moreover, because of the variations in the success rates for CAR-T cell therapies, a majority of patients will therefore still require salvage therapies and palliative care [12]. Although the short-term cost burden is likely to be relatively small given that current and emerging products are mostly indicated for rare diseases, it is the cumulative effects of multiple products on the horizon targeting more prevalent diseases that raised financial concern for the longer term [13]. Payers are facing an opportunity cost dilemma when choosing between paying for gene therapies and chronic treatments within their constrained budget [14].

7.4 THE IMPLEMENTATION OF MANAGED ENTRY AGREEMENTS

As cell and gene therapies generally have a high price tag, but immature evidence to demonstrate their effectiveness and safety, HTA bodies and payers are struggling to find effective strategies to endorse this class of innovative products. Managed entry agreements (MEAs) have been used by payers and the industry as a solution to this challenge [15]. In general, MEAs could be classified into three categories: (1) financial-based payment, such as price-volume agreement, discount, utilisation cap, and rebates, (2) performance-based payment, such as coverage with evidence development, performance-based payment, risk-sharing agreements, and instalment payment spreading the cost over several years whereby the payment is linked to the sustained treatment benefits, and service-based agreement ref paper with Monique in ViH. For example, performance-based MEAs are used by payers to mitigate affordability, address value uncertainty without limiting patient access to cell and gene therapies [16].

According to the 2019 OECD Health Working Paper: Performance-based MEAs for new medicines in OECD countries and European Union (EU) member states [17], the use of MEAs has increased over time. MEAs were being or had been used in at least 28 of 41 countries that are members of the OECD and/or the EU. Some countries have established specific legislation or policies governing MEAs, such as in Belgium, the Czech Republic, Lithuania, Norway, Portugal, and Slovakia. While there was no specific legislation, countries such as France or Italy also had in place defined processes for MEAs based on laws related to the coverage of medicines. In Germany, contracts between sickness funds and pharmaceutical firms that make pricing subject to specified conditions are permissible by legislation related to the pharmaceutical market, and it is up to individual sickness funds to decide if and how to use them. The implementations of MEAs for gene therapies in major European countries are detailed in Table 7.1.

7.4.1 Yescarta® (Axicabtagene Ciloleucel)

7.4.1.1 UK

In January 2019, NICE considered that the evidence from a single-arm study is uncertain because of limited follow-up and the lack of direct data comparing axicabtagene ciloleucel with salvage chemotherapy. Moreover, there is also not enough evidence to determine the costs of treating side effects. As a result, Yescarta was not recommended for routine use in the UK, while it was reimbursed through the Cancer Drugs Fund (CDF), which is a cancer-specific funding source that enables access to drugs on the condition of gathering more evidence on the real-world effectiveness (i.e., coverage with

Table 7.1 Managed Entry Agreements for Gene Therapies in the Europe Market

	United Kingdom	France	Germany	Italy	Spain
Yescarta®	CDF with commercial agreements	Coverage with evidence development	Coverage with evidence development	Staged, outcome-based payment	Staged, outcome-based payment
Kymriah®	CDF with commercial agreements	Coverage with evidence development	Coverage with evidence development	Staged, outcome-based payment	Staged, outcome-based payment
Zynteglo®	N/A	Coverage with evidence development	Coverage with evidence development	N/A	N/A
Zolgensma®	Simple discount; patient access scheme	Coverage with evidence development	Evaluation suspended	N/A	N/A
Tecartus®	CDF with managed access scheme	Coverage with evidence development	Coverage with evidence development	N/A	N/A
Libmeldy®	N/A	Coverage with evidence development	Coverage with evidence development	N/A	N/A

Note: CDF – Cancer Drug Fund; N/A – Not available.

evidence development). Additionally, confidential discounts must be given to National Health Services (NHS), coupled with a requirement to collect and submit additional data for the purposes of a future price reassessment. The evaluation committee is expecting the long-term follow-up data from ZUMA-1 to resolve the uncertainties on the progression free, post-progression, and overall survival data, and to gain a more complete understanding of the treatment for B-cell aplasia including the cost and use of IVIG. The data on using axicabtagene ciloleucel in the NHS would be collected using the Systemic Anti-Cancer Therapy (SACT) dataset, which would more accurately reflect the costs and benefits of its use in clinical practice.

7.4.1.2 France

Before the central approval of the EU, Yescarta was available in France through the early access program called "Temporary Authorisation for Use" (Autorisation Temporaire d'Utilisation, [ATU]). During the ATU/post-ATU period, the manufacturers are allowed to set the drug price freely; however, the pricing committee sets a maximum price per unit. In addition, drugs with annual pre-tax revenue exceeding €30 million under the ATU/post-ATU period are subject to spending caps, above which manufacturers are liable to pay rebates. It is worthy to mention that the most recent French Social Security law of 14 December 2020 has proposed several changes to the existing ATU system. The ATU program has now been grouped into two: an early access authorisation program (Accès précoce aux médicaments, EAP) which includes the cohort ATU (a group of patients) and a compassionate access program (Accès compassionnel aux médicaments, CAP) which includes nominative nATU (single patient), among others. There are some changes to the new reform: (1) Manufacturers will now be required to make all EAP requests to the HTA body (Haute Autorité de Santé, HAS), not to the regulatory agency (Agence nationale de sécurité du médicament et des produits de santé, ANSM); (2) the introduction of the presumption of innovation for the products, compared with the most clinically relevant comparator; (3) CAP criteria are now stricter as the manufacturer needs to demonstrate that the efficacy and safety of a given product is "strongly presumed," as opposed to "considered" to be favourable under the old system; (4) The manufacturer must not only follow a well-established clinical protocol and collect data during the period required by HAS but also needs to demonstrate that trials have started. As a result, the standards for evidence requirements of early access programs become higher, and fewer patients will be able to have access to highly needed products.

In December 2018, HAS assessed that Yescarta® to have an important actual clinical benefit (service médical rendu [SMR]) and a moderate improvement in added clinical benefit over existing therapies (amélioration du service médical rendu [ASMR]: III) for both indications. Due to the high uncertainty around the effectiveness and the ancillary implications of delivering the CAR-T cell therapies (e.g., cost of hospitalisation, etc.), HAS will undertake annual HTAs, reassessing ASMR using the data collected in the CAR-T specific registry, as well as any new data collected from the pivotal studies. In April 2021, the updated HAS report was published, in which HAS maintained their favourable opinions towards the assessment (SMR: important, ASMR: III) and reimbursement of Yescarta®. Moreover, HAS calls for the participation of all qualified centres in the DESCAR-T register to obtain comprehensive and high-quality observational data that could be valuable to address the evidence uncertainties in the next re-assessment.

7.4.1.3 Germany

Manufacturers of new drugs launching in Germany are allowed to price the drugs freely for the first 12 months after launch, during which an HTA is undertaken, and a reimbursement recommendation will be made by the Federal Joint Committee (G-BA). The added benefit of Yescarta® was considered as "proven" due to its orphan drug status, while the extent of the added benefit of Yescarta® was evaluated as "non-quantifiable." Due to the evidence limitations, the decision made by the G-BA will be valid until 15 May 2022, at which time the G-BA will update the benefit assessment of Yescarta® based on new evidence, which could be generated from an ongoing clinical trial or prospective comparative studies beyond the pivotal trial. Outcome-based payment schemes have historically not been widely adopted in Germany, while the recent gene therapy launches have seen a shift towards an increase in the application. Gilead signed outcome-based payment schemes based on individual patient's outcomes for Yescarta with several health insurers including two umbrella groups of insurers: VDEK and GWQ ServicePlus.

7.4.1.4 Italy

The launch of Yescarta and Kymriah has seen a shift of Italy's traditional outcome-based payment towards a staged payment scheme. For Yescarta®, Gilead will be paid in three instalments: the first payment for Yescarta® is scheduled at 180 days after infusion, the second payment at 270 days, and the final payment at 365 days. In such a case, Gilead may bear the risk of receiving no payments for patients who showed no treatment benefits in 180 days. Data will be collected through the Italian Medicines Agency (AIFA) registry. These contractual arrangements for Yescarta are valid for 18 months, from 4 November 2019.

7.4.1.5 Spain

Yescarta® was approved for reimbursement with two outcome-based staged payment schemes linked to patient-level survival evidence in July 2019, reportedly the first payment of €118,000 and the second payment of €209,000. The post-launch evidence for Yescarta will be collected through a new data collection and management system for the Spanish NHS called Valtermed (an abbreviation for Sistema de Información para determinar el Valor Terapéutico en la Práctica Clínica Real de los Medicamentos de Alto Impacto Sanitario y Económico en el SNS2). The Valtermed registry system is designed to collect real-world clinical data through a web-based tool to reduce the uncertainty in treatment benefits associated with new therapies in real-world practice.

7.4.2 Kymriah® (Tisagenlecleucel)

7.4.2.1 UK

Similar to Yescarta®, Kymriah® was accepted in the Cancer Drug Fund on the condition that the managed access agreement is followed and the requirements for future data collection for re-assessments are met. NICE considers that a longer follow-up study is needed and there are no data directly comparing tisagenlecleucel with salvage chemotherapy. Collecting more data on progression-free survival, overall survival, and immunoglobulin usage will reduce the uncertainty in the evidence.

7.4.2.2 France

Similar to Yescarta®, Kymriah® was used on the ATU basis before the central market authorisation was granted. In December 2018, Kymriah® was considered to have an important actual clinical benefit (SMR: important) and a moderate improvement in added benefit over available therapies (ASMR: III) for the treatment of B-cell ALL [18]. For the diffuse large B-cell lymphoma (DLBCL) indication, Kymriah® was assessed having an important actual clinical benefit (SMR: important), but a minor improvement in added benefit over the comparator therapies (ASMR: IV). In the renewed HAS assessment released in April 2021, HAS retained the previous assessment that Kymriah® is of moderate and minor added clinical benefits for B-cell ALL and DLBCL, respectively [19]. More evidence collected in the DESCAR-T register will be requested for the re-assessment.

7.4.2.3 Germany

In March 2019, Kymriah® was assessed with a "non-quantifiable" added clinical benefits for both indications [20]. Novartis signed an outcome-based contract with VDEK and GWQ ServicePlus whereby the companies will provide rebates to the health insurers for patients dying after treatment with Kymriah® within a period of time. In the renewed G-BA assessment released in September 2020 [21], G-BA retained their previous assessment that the added clinical benefit of Kymriah® is "non-quantifiable." G-BA requested further data must be submitted, and the current decision will be valid until September 2023.

7.4.2.4 Italy

Similar to Yescarta®, Kymriah® was reimbursed through an outcome-based payment in Italy, whereby the payment will be made at three time points (assuming that the patient sustains the agreed health benefit): The first at the time of infusion, the second after six months, and the third after 12 months. Future data will be collected through the AIFA registry.

7.4.2.5 Spain

Similar to Yescarta®, Kymriah® is reimbursed in the Spanish NHS through outcomes-based, staged payments using the data collected through the Valtermed system: one at the time of infusion (reported to be 52% of the total €320,000), and a second payment at 18 months (reportedly the remaining 48%), provided that the patient has sustained the treatment responses.

7.4.3 Zynteglo® (Betibeglogene Autotemcel)

7.4.3.1 France

In March 2020, the HAS considered that Zynteglo was of "important" clinical benefit for patients aged 12–35 years, in the treatment of patients with transfusion-dependent ß-thalassemia (TDT), who do not have an $\beta 0/\beta 0$ genotype, eligible for hematopoietic stem cell transplantation (HSC), but who do not have a related HLA (human leukocyte antigen) donor available. In the patient population, HAS evaluated that Zynteglo of "moderate" added benefits (ASMR III). In contrast, the clinical benefit of Zynteglo is considered insufficient in patients aged 35 years and over. The Commission will re-evaluate Zynteglo within a maximum period of 3 years based on the new data collected from the patient registry and the results of pivotal studies requested by the EMA [22].

7.4.3.2 Germany

In May 2020, Zynteglo was assessed with a hint for a "non-quantifiable" additional benefit considering the evidence limitation. A renewed benefit assessment is to be based on the final findings of the studies HGB-207 and HGB-212, as well as on the 5-year follow-up data of the study LTF-303. In line with the conditions imposed by the EMA, a renewed benefit assessment should also include data from a registry study on the long-term safety and efficacy of betibeglogene autotemcel in patients aged 12 years and older with transfusion-dependent β thalassaemia (TDT) who do not have a $\beta 0/\beta 0$ genotype. The registry study should compare data from the product registry REG-501 with data from patients treated with transfusions from an established European registry. For this purpose, the G-BA considers a limitation of the resolution until 15 May 2025 to be appropriate [23].

7.4.4 Zolgensma® (Onasemnogene Abeparvovec)

7.4.4.1 UK

Zolgensma was evaluated through the Highly Specialised Technology (HST) program in the UK. The NICE evaluation for Zolgensma was released in July 2021, which includes three recommendations:

- Onasemnogene abeparvovec is recommended as an option for treating 5q spinal muscular atrophy (SMA) with a bi-allelic mutation in the SMN1 gene and a clinical diagnosis of type 1 SMA in babies, only if:
  - they are 6 months or younger, or
  - they are aged 7–12 months, and their treatment is agreed upon by the national multidisciplinary team. It is only recommended for these groups if:
  - permanent ventilation for more than 16 hours/day or a tracheostomy is not needed.
- For babies aged 7–12 months, the national multidisciplinary team should develop auditable criteria to enable onasemnogene abeparvovec to be allocated to babies in whom treatment will give them at least a 70% chance of being able to sit independently.
- Onasemnogene abeparvovec is recommended as an option for treating pre-symptomatic 5q SMA with a bi-allelic mutation in the SMN1 gene and up to three copies of the SMN2 gene.

A simple discount patient access scheme is applied for recommendation 1 and recommendation 2, while a managed access agreement, which includes a patient access scheme is applied for recommendation 3. Recommendation 1 and recommendation 2 will be valid until 2024. The re-assessment for recommendation 3 will be performed after 1 year of the launch using data from an ongoing trial, combined with data from NHS routine population-wide datasets.

7.4.4.2 France

The HAS assessment of Zolgensma was released in December 2020 [24]. The Committee deems that the clinical benefit of Zolgensma (onasemnogene abeparvovec) is:

- Substantial in the treatment of patients with 5q SMA with a bi-allelic mutation in the SMN1 gene and a clinical diagnosis of SMA Type 1 and 2 or pre-symptomatic patients, with up to three copies of the SMN2 gene.
- Insufficient to justify public funding cover in all other clinical situations. In the treatment of patients with 5q SMA with a bi-allelic mutation in the SMN1 gene and a clinical diagnosis of SMA Type 3.

The Committee deems that the clinical benefit of Zolgensma (onasemnogene abeparvovec) is:

- Moderate (ASMR III), in the same way as Spinraza (nusinersen), in the care pathway of pre-symptomatic patients with a genetic diagnosis of SMA with a bi-allelic mutation in the SMN1 gene and one to two copies of the SMN2 gene.
- No (ASMR V) in the care pathway, excluding Spinraza (nusinersen), of pre-symptomatic patients with a genetic diagnosis of SMA with a bi-allelic mutation in the SMN1 gene and three copies of the SMN2 gene.

More data will be collected from the RESTORE registry as required by EMA. Additionally, the patients treated with Zolgensma should also be included in the French national registries for SMA France, which is put in place for nusinersen previously. The inclusion of SMA patients in the national registry is important to further re-assessments of these two treatments. This collection of data should make it possible to document the characteristics of the disease by type, assessment of motor, respiratory and neurocognitive functions, QoL, tolerance, mortality, and the treatment strategy of (symptomatic and pre-symptomatic) SMA patients. The Commission expects to obtain the data from these registers within a maximum period of 5 years. In addition, it will re-evaluate Zolgensma (onasemnogene abeparvovec) within 3 years based on data from ongoing studies (clinical trials and ATU) and data from the register [25].

7.4.4.3 Germany

Orphan drugs in Germany could benefit from an automatically "proven" added benefit from the G-BA at the time of approval, without the need for a comparative assessment against an appropriate comparator therapy (ACT). However, if the sales values exceed €50 million per year, an ACT must be defined, and a full benefit assessment must be undertaken. During the first 6 months of sales in Germany, Zolgensma exceeded this threshold. The "law for more safety in the supply of pharmaceuticals (GSAV)" was passed by the German parliament and took place in June 2019. This enables the G-BA to require additional real-world evidence (RWE) following the initial assessment for products with conditional approval and orphan drugs (given due to limited evidence). In July 2020, during the review of Zolgensma, G-BA explicitly specified the rules for postlaunch evidence collection for orphan drugs approved, with considerable evidence of uncertainties. Zolgensma represents the first orphan drug that is obligatory to collect the RWE in Germany (February 2021).

Once determined that additional data collection is necessary, the IQWIG, on behalf of the G-BA, will draft concepts, including the detailed requirements of type, duration, scope, methods of data collection, and the patient-relevant endpoints to be considered as outcomes. In addition, the G-BA stated that at least every 18 months, they will evaluate whether data collection is being carried out or can no longer be carried out, whether it will provide sufficient evidence for a renewed benefit assessment, or whether there is a need for adjustments to the provisions of the decision [26]. To meet the G-BA's requirement, Novartis will have to perform a registry-based study to include up to 500 children and effectiveness compared with Biogen's nusinersen (using SMArtCARE registry). The G-BA reassessment of Zolgensma is planned at the latest in summer 2027.

Outcome-based payment with rebates based on several patient-relevant outcomes (not disclosed) was not signed with sick fund insurers (GWQ ServicePlus AG) potentially using data from the EU registry. In contrast to previous contracts with a similar innovative reimbursement model, AveXis assumes the risk of repaying up to 100% of the drug costs in a staggered manner in the event of a contract [27].

7.4.4.4 Italy

In March 2021, Zolgensma was assessed with a score "important" for patients weighing up to 13.5 kg and clinical diagnosis of SMA type 1 and onset of symptoms during the first six months of life or genetic diagnosis of SMA type 1 (biallelic mutation in the SMN1 gene and up to two copies of the SMN2 gene) in Italy [28]. An outcome-based payment deal was signed between Norvartis and the NHS, in which the payment will be made in five time points at baseline, 12 months, 36 months, and 48 months based on achieved and sustained outcomes.

7.4.5 Tecartus® (Brexucabtagene Autoleucel)

7.4.5.1 UK

In February 2021, Tecartus was recommended to be used under the Cancer Drug Fund on the condition that the managed access agreement was followed. Tecartus meets NICE's criteria to be considered a life-extending treatment at the end of life, while there are evidence uncertainties due to the short follow-up, a small number of patients, long-term treatment benefits and lack of comparative evidence against the most common alternative treatment.

7.4.5.2 France

The HAS assessment of Tecartus was released in May 2021. The Committee deems that Tecartus (onasemnogene abeparvovec) is of "substantial" clinical benefit and "moderate" clinical added benefits considering that (1) relevant short-term efficacy data from a non-comparative phase II study (ZUMA-2) in terms of complete response (approximately 60% of the intention-to-treat [ITT] population) and overall survival (69% of living patients with a theoretical median follow-up of 16.8 months), (2) uncertainties about the magnitude of effect due to the lack of direct comparison with existing management and the limitations of indirect comparisons made, (3) uncertainties about the maintenance of clinical efficacy in the longer term and in particular about the achievement of cures in patients in long-term remission, (4) significant short-term toxicity and lack of long-term safety data.

Given the uncertainties about efficiency, tolerability, and the complexity of the treatment (e.g., patient eligibility, leukapheresis, reinjection of CAR-T cells, and post-infusion monitoring), the Commission requests that more data must be collected in the recipients of Tecartus from the follow-up of the ZUMA-2 study, the final report of the ATU, and evidence from EBMT register requested by the EMA [29]. Additionally, HAS required the establishment of a comprehensive registry for all patients eligible for Tecartus, including those as part of the post-ATU program, based on the French DESCAR-T study common to CAR-T cell drugs. As with previous CAR-T cell medicines evaluated by the Commission, it specifies that the collection of data must concern all patients eligible for the drug in France and do not concern only patients who actually received the treatment. This data should support short- and long-term efficacy and safety and identify predictive factors of response to treatment. This data should also make it possible to describe in real conditions of use, such as the characteristics of patients eligible for treatment and those of patients actually treated, their treatment history, the characteristics of the disease at the time of eligibility and at reinjection, the conditions of use and therapeutic strategies put in place before and after the reinjection, the persistence of CAR T, the time between the failure of the previous treatment line and the apheresis as well as the delay until administration to the patient, and the causes of treatment failure and subsequent management. The reassessment will be performed after 2 years of issuing the first assessment with the additional evidence collected in the post-launch stage.

7.4.5.3 Germany

In August 2021, the extent of the added value of Tecartus was assessed as "non-quantifiable" due to the limited evidence. Tecartus will be suspectable to quality assurance minimum requirements for CAR-T cell therapies (e.g., Yescarta and Kymriah), which concern the infrastructure of the medical facility, the existing nursing, and specialist treatment competence as well as the experience with these novel cell therapies. The aim is a secure indication, as the benefit-risk assessment can look different for each patient, as well as optimal complication-free treatment and follow-up. Further quality assurance requirements concern the preparation and administration of car-T cells, as incorrect handling of the products can limit the possible therapeutic success [30].

7.4.6 Libmeldy® (Autologous CD34+ Cells Encoding ARSA Gene)

7.4.6.1 France

In April 2021, HAS assessed that Libmeldy is of "important" clinical benefit only in asymptomatic children without clinical manifestation of the disease, whether in terms of motor, cognitive, and/or behavioural impairment, or with the late infantile form (manifested before 30 months), or early juvenile form (manifested between 30 months and 6 years inclusive) of metachromatic leukodystrophy. In this population, the added clinical benefit of Libmeldy is assessed as "moderate" (ASMR III). In contrast, HAS that the clinical benefit of Libmeldy is insufficient for symptomatic children, with early clinical manifestations of the disease, which retained the ability to walk independently and before the onset of cognitive decline, with the early juvenile form (manifested between 30 months and 6 years inclusive).

The Commission requested that more evidence should be collected from the final reports of ongoing study 201222 and study 205756, patient follow-up data included in early access programs, and the final report of the long-term observational MLD study. In addition, the Commission expects to have data to describe all patients eligible for Libmeldy in France, either actually treated or not – the characteristics of patients eligible for LIBMELDY, such as the monitoring of motor development and neurocognitive of children, QoL, possible predictors of response to treatment, and tolerance [31]. A reassessment will be conducted in 3 years after the additional data requested above is available.

Moreover, in October 2021, Libmeldy was included in the new early access program-Autorisation d'accès précoce (AAP) based on the assessment that Libmeldy is presumed to be innovative (in the light of clinically relevant comparators) in particular because it is a new modality of disease management bringing a substantial change to patients in terms of efficacy, safety, and care pathway.

7.4.6.2 Germany

In November 2021, G-BA assessed the additional benefit of Libmeldy (atidarsagen autotemcel) is

- A hint of considerable for children with late infantile (LI) or early juvenile (EJ) forms of metachromatic leukodystrophy (MLD) without clinical manifestations of the diseases.
- A hint of non-quantifiable for children with the EJ form of metachromatic leukodystrophy with early clinical manifestations of the disease who still have the ability to walk independently, before the onset of cognitive decline.

The EMA has requested the submission of the final study report of the ongoing pivotal study 201222 with the active ingredient Libmeldy (atidarsagen autotemcel) for 31 March 2024. It is expected that more evidence concerning overall survival and sustainability of the effects in the endpoint category morbidity for two patient populations (patients with and without clinical manifestations of the disease) will be available by then, which will enable a more reliable assessment of the added benefits of Libmeldy. Therefore, the initial

G-BA assessment will be valid until March 2024, then a reassessment will be initiated with new data submitted.

7.5 CHALLENGES AND RECOMMENDATIONS IN THE IMPLEMENTATION OF MEA

In general, while financial MEAs have proliferated, the use of performance-based MEAs remains more limited. This is likely to be explained that these agreements are resource intensive to implement and require good IT systems with electronic clinical records linked to reimbursement systems to be successfully enacted [32]. The implementation of MEA highly relies on the collection of more comprehensive evidence in the post-launch phase to address the evidence gap in the initial assessment. In most countries, data was collected through a range of sources, such as the existing health data or insurance data), combined with those ongoing clinical studies will be completed. For CAR-T cell therapies including Yescarta and Kymriah, some countries (e.g., France) required that the data will also be collected with existing chemotherapy datasets and national registries for bone marrow transplant or CAR-T products. Italy and Spain have a national web-based platform for which bespoke data collection requirements are created for each medicine/therapeutic indication. However, in Italy, the national platform is not linked to the hospital reimbursement system, and thus duplicate data entry is required by the prescribers. In Spain, the platform is being established and was not fully functional for CAR-T cell therapies [33]. Moreover, there are growing concerns surrounding the fulfilment of post-market obligations. It was indicated that most post-market studies required by regulators were completed with a substantial delay and showed methodological discrepancies over time [34, 35]. Information relating to randomisation methods, comparator types, outcomes, and patient numbers were not sufficiently reported [36, 37]. One of the most important reasons is the lack of incentives for manufacturers to collect post-launch evidence once products are on the reimbursement list. This is linked to the problem of delisting a new technology, once patients and clinicians have access to, and are familiar with it [38].

As a result, it is unclear to know how MEAs achieved their purpose in real practice. Although MEAs provided the short-term advantage of covering new medicines at lower confidential prices, they may have limited impacts on reducing uncertainty around the comparative effectiveness and cost effectiveness of the investigated products [17]. Neyt et al. evaluated the use of MEAs in Belgium and concluded that while MEAs have a positive impact on early access to products, there are some risks in its implementation. Recommendations were proposed to ensure that MEAs could realise its promises to promote more informed decision-making on resource allocations: (1) The type of MEA (financial-based agreement, performance-linked agreement, and coverage with evidence development) should be customised to the evidence uncertainty that needs to be addressed; (2) transparent information should be provided to patients and the prescribing physician about the "temporary" nature of the funding and the possibility of delisting; (3) MEAs should not be the standard, but it should only be considered as options for well-considered cases, such as innovative products to address diseases with high unmet medical needs; (4) more international collaboration is needed to increase the information sharing, which could be helpful to make better MEAs plans as well as to inform (public) price negotiations.

7.6 CONCLUSION

With the growing number of cell and gene therapies that will be launched soon, payers will face mounting pressures to find effective strategies to mitigate the affordability challenges.

Considering the limited experience in answering how the out-come-based MEAs will impact market access of cell and gene therapies approved in the past years, more evidence is awaited to examine whether outcome-based MEAs could be sustainable solutions to satisfy the expectations of each party. Countries should establish a clear policy framework to regulate their imple-mentations (e.g., eligibility criteria) and strengthen the monitor-ing system to ensure that the post-launch scientific obligations are adequately followed. With the lack of clarity and strict mea-surements, the effects of outcome-based MEAs in accelerating patient access to life-saving therapies will be undermined.

REFERENCES

1. Qiu T, Hanna E, Dabbous M, Borislav B, Toumi M. Regenerative Medicine Regulatory Policies: A Systematic Review and International Comparison. Health Policy. 2020;124(7):701–13.
2. Conditional marketing authorization – Report on ten years of experience at the European Medicines Agency. Available from: https://www.ema.europa.eu/en/documents/report/con-ditional-marketing-authorisation-report-ten-years-experience-european-medicines-agency_en.pdf
3. Orphan Medicines Figures – 2000–2020. Available from: https://www.ema.europa.eu/en/documents/other/orphan-medicines-figures-2000-2020_en.pdf
4. Abrahamyan L, Feldman BM, Tomlinson G, et al. Alternative Designs for Clinical Trials in Rare Diseases. Am. J. Med. Genet. C Semin. Med. Genet. 2016;172(4):313–31.
5. Coyle D, Durand-Zaleski I, Farrington J, et al. HTA Methodology and Value Frameworks for Evaluation and Policy Making for Cell and Gene Therapies. Eur. J. Health Econ. 2020;21(9):1421–37.
6. Drummond MF, Neumann PJ, Sullivan SD, et al. Analytic Considerations in Applying a General Economic Evaluation Reference Case to Gene Therapy. Value Health 2019;22(6):661–68.
7. Carr DR, Bradshaw SE. Gene Therapies: The Challenge of Super-High-Cost Treatments and How to Pay for Them. Regen. Med. 2016;11(4):381–93.

8. Faulkner E, Spinner DS, Ringo M, Carroll M. Are Global Health Systems Ready for Transformative Therapies? Value Health 2019;22(6):627–41.

9. Towse A, Fenwick E. Uncertainty and Cures: Discontinuation, Irreversibility, and Outcomes-Based Payments: What Is Different About a One-Off Treatment? Value Health 2019;22(6):677–83.

10. Prasad V. Immunotherapy: Tisagenlecleucel – The First Approved CAR-T-Cell Therapy: Implications for Payers and Policy Makers. Nat Rev Clin Oncol. 2018;15(1):11–12.

11. Roth JA, Sullivan SD, Lin VW, et al. Cost-Effectiveness of Axicabtagene Ciloleucel for Adult Patients With Relapsed or Refractory Large B-Cell Lymphoma in the United States. J. Med. Econ. 2018;21(12):1238–45.

12. Champion AR, Lewis S, Davies S, Hughes DA. Managing Access to Advanced Therapy Medicinal Products: Challenges for NHS Wales. Br. J. Clin. Pharmacol. 2020. doi:10.1111/bcp.14286

13. Barlow JF, Yang M, Teagarden JR. Are Payers Ready, Willing, and Able to Provide Access to New Durable Gene Therapies? Value Health 2019;22(6):642–7.

14. Patel N, Farid SS, Morris S. How Should We Evaluate the Cost-Effectiveness of CAR T-Cell Therapies? Health Policy Technol. 2020. doi:10.1016/j.hlpt.2020.03.002

15. Dabbous M, Chachoua L, Caban A, Toumi M. Managed Entry Agreements: Policy Analysis From the European Perspective. Value Health 2020;23(4):425–33.

16. Michelsen S, Nachi S, Van Dyck W, Simoens S, Huys I. Barriers and Opportunities for Implementation of Outcome-Based Spread Payments for High-Cost, One-Shot Curative Therapies. Front. Pharmacol. 2020;11:594446.

17. OECD Health Working Paper No. 115: Performance-Based Managed Entry Agreements for New Medicines in OECD Countries and EU Member States. Available from: https://www.oecd.org/health/health-systems/HWP-115-MEAs.pdf

18. KYMRIAH (tisagenlecleucel), CAR T anti-CD19 (DLBCL). Available from: https://www.has-sante.fr/jcms/c_2891692/fr/kymriah-tisagenlecleucel-car-t-anti-cd19-ldgcb

19. Inokuma Y. Pharmacovigilance of Regenerative Medicine Under the Amended Pharmaceutical Affairs Act in Japan. Drug Saf. 2017;40(6):475–482.

20. FDA announces comprehensive regenerative medicine policy framework. Available from: https://www.fda.gov/NewsEvents/Newsroom/PressAnnouncements/ucm585345.htm

21. Adminstration FaD. Expedited_Programs_Regenerative_Medicine_Therapies_Serious_Conditions_Final.pdf. 2019.

22. Tigerstrom BV. Revising the Regulation of Stem Cell-Based Therapies: Critical Assessment of Potential Models. Food Drug Law J. 2015;70:315–37.

23. Administration FaD. Regulatory Consideration for Human Cells, Tissues and Cellular and Tissue- based Products: Minimal Manipulation and Homologous Use, 2017.

24. Union E. Regulation (EC) No 1394/2007 of the European Parliament and of the Council of 13 November 2007 – On Advanced Therapy Medicinal Products and Amending Directive 2001/83/EC and Regulation (EC) No 726/2004. Official Journal of European Union.

25. Guidelines relevant for advanced therapy medicinal products. 2019. Available from: https://www.ema.europa.eu/en/human-regulatory/research-development/advanced-therapies/guidelines-relevant-advanced-therapy-medicinal-products

26. Qiu T, Wang Y, Liang S, Han R, Toumi M. The Impact of COVID-19 on the Cell and Gene Therapies Industry: Disruptions, Opportunities, and Future Prospects. Drug Discovery Today 2021. doi:10.1016/j.drudis.2021.04.020 S1359-6446(1321)00207-00205

27. Ronfard V, Vertes AA, May MH, Dupraz A, Van Dyke ME, Bayon Y. Evaluating the Past, Present, and Future of Regenerative Medicine: A Global View. Tissue Eng. Part B Rev. 2017;23(2):199–210.

28. Gozzo L, Romano GL, Romano F, et al. Health Technology Assessment of Advanced Therapy Medicinal Products: Comparison Among 3 European Countries. Front Pharmacol. 2021;12:755052.

29. Support for advanced-therapy developers. Available from: https://www.ema.europa.eu/en/human-regulatory/research-development/advanced-therapies/support-advanced-therapy-developers

30. Agency EM. Procedural Advice on the Evaluation of Advanced Therapy Medicinal Product in Accordance with Article 8 of Regulation (EC) No 1394/2007, 2018.

31. Cellular & Gene Therapy Products. Available from: https://www.fda.gov/BiologicsBloodVaccines/CellularGeneTherapyProducts/default.htm

32. Ferrario A, Araja D, Bochenek T, et al. The Implementation of Managed Entry Agreements in Central and Eastern Europe: Findings and Implications. Pharmacoeconomics 2017;35(12):1271–85.

33. Facey K-O, Espin J-O, Kent E-OX, et al. Implementing Outcomes-Based Managed Entry Agreements for Rare Disease Treatments: Nusinersen and Tisagenlecleucel. (Electronic). 1179–2027.

34. Banzi R, Gerardi C, Bertele V, Garattini S. Approvals of Drugs With Uncertain Benefit-Risk Profiles in Europe. Eur. J. Intern. Med. 2015;26(8):572–84.

35. Hoekman J, Klamer TT, Mantel-Teeuwisse AK, Leufkens HG, De Bruin ML. Characteristics and Follow-up of Postmarketing Studies of Conditionally Authorized Medicines in the EU. Br. J. Clin. Pharmacol. 2016;82(1):213–26.

36. Wallach JD, Egilman AC, Dhruva SS, et al. Postmarket Studies Required by the US Food and Drug Administration for New Drugs and Biologics Approved between 2009 and 2012: Cross Sectional Analysis. BMJ 2018;361:k2031.

37. Makady A, Van Veelen A, De Boer A, Hillege H, Klungel OH, Goettsch W. Implementing Managed Entry Agreements in Practice: The Dutch Reality Check. Health Policy. 2019;123(3):267–74.

38. Neyt M, Gerkens S, San Miguel L, Vinck I, Thiry N, Cleemput I. An Evaluation of Managed Entry Agreements in Belgium: A System With Threats and (high) Potential If Properly Applied. Health Policy. 2020;124(9):959–64.

Index

Note: Locators in *italics* represent figures and **bold** indicate tables in the text.

Printed in the United States
by Baker & Taylor Publisher Services

Printed in the United States
by Baker & Taylor Publisher Services